ISBN 978-1-333-71614-1
PIBN 10538599

English
Français
Deutsche
Italiano
Español
Português

www.forgottenbooks.com

Mythology Photography **Fiction**
Fishing Christianity **Art** Cooking
Essays Buddhism Freemasonry
Medicine **Biology** Music **Ancient
Egypt** Evolution Carpentry Physics
Dance Geology **Mathematics** Fitness
Shakespeare **Folklore** Yoga Marketing
Confidence Immortality Biographies
Poetry **Psychology** Witchcraft
Electronics Chemistry History **Law**
Accounting **Philosophy** Anthropology
Alchemy Drama Quantum Mechanics
Atheism Sexual Health **Ancient History**
Entrepreneurship Languages Sport
Paleontology Needlework Islam
Metaphysics Investment Archaeology
Parenting Statistics Criminology
Motivational

PORTOLAN CHARTS

THEIR ORIGIN AND CHARACTERISTICS

WITH A DESCRIPTIVE LIST OF THOSE BELONGING TO

THE HISPANIC SOCIETY OF AMERICA

BY

EDWARD LUTHER STEVENSON, PH.D.

NEW YORK

1911

The Knickerbocker Press, New York

CONTENTS

REPRODUCTIONS

PORTOLAN CHARTS

AMONG the geographical records of earlier centuries which have come down to us, none are more interesting than the portolan charts which were drawn during the years fittingly designated the period of great geographical discoveries. They attract and hold the attention by reason of their artistic features, as well as by their remarkable approach to scientific accuracy for so early a period.

To the cloister maps of the middle ages they present a marked contrast. The former strikingly exhibit the influence of ecclesiastical and classical tradition. In general, they are far from truthful in their presentation of the geographical features of the earth. Though highly interesting as reflecting geographical notions of the time in which they were drawn, they possess little value as scientific maps.

Portolan charts are based upon careful and what may be called scientific observations. It is only in recent times that there has been an improvement in the charting of the region to which most of them pertain, that is, the Mediterranean and the Atlantic coast in varying extent to the north and the south of Gibraltar. They too exhibit the geographical interests of the period to which they belong. They are the creations of seamen, navigators, explorers, chart-makers who were leaders in the expansion of geographical knowledge which opened the New World region of Africa, of India, and of America.

This brief word concerning the origin, character, and general significance of portolan charts, the first modern scientific maps, is presented as an introduction to a descriptive list of the numerous originals belonging to The Hispanic Society of America. An inquiry into the history of portolan charts which have been preserved to our day leads immediately to a query concerning their origin. None of those extant are known to have been drawn prior to the year 1300, and the oldest example bearing date and signature is that of Pietro Visconte of the year 1311. Nordenskiöld thinks that the normal portolan chart, as he chooses to call it, that is, the one which served as a sort of original pattern, must have been constructed sometime during the thirteenth century, from numerous coast sketches such, for example, as those which may be found in a cosmographic poem by Leonardo Dati, bearing the title *La sfera*. The arguments in support of the assumption seem reasonable, yet the fact remains that no dated portolan chart of that century, as has been stated above, is known; neither are such sketches known as those to which Nordenskiöld refers, antedating the fourteenth century. An interesting record, however, is that to be found in a work by Guillaume de Nangis describing the crusade of King Louis IX. in 1270, noting that the King's ships had sea charts on board. In the voyage from Aiguesmortes to Cagliari, the port selected for the rendezvous of the ships making up the expedition, they were overtaken by a storm, and at the end of the sixth day, as Cagliari had not yet been reached, the King expressed a wish to know the exact location of his ship. The pilots, we are told, brought to him their charts, and showed to him that the port was not far distant.

Theobald Fischer has advanced the theory that portolan charts have a Byzantine origin, and Fiorini holds that Italian navigators, not long after 1000 A. D., learned from the Greeks of Constantinople how to make and how to use charts which were founded on drawings and measurements, and that in succeeding years they gradually improved them. Again the fact confronts us that no portolan chart of Byzantine or Greek origin is known, nor is the evidence of such eastern influence traceable in existing charts.

The first thousand years and more of the Christian era have left us none of the sailors' charts which may have been employed during those centuries.

Ptolemy alone of the ancient writers alludes to the charts of seamen, and one might conclude from his references that such as he had in mind were not unlike the portolan charts which we have here under consideration. But all these too are lost.

As there appears to be a relationship existing between the ancient periplus, the Italian portolan, and the portolan chart of the period of discovery,—which chart at first was doubtless regarded as a very useful addition to the portolan, coming in time to supplant it as the knowledge of seamanship expanded,—a more extended reference to the character of the periplus and of the portolan will fittingly introduce us to the portolan chart.

The Greeks used the word περίπλους to designate a course or harbor book, literally a sailing around, a circumnavigation. It was not applied to a sea chart or to a collection of sea charts. The Italian word *portolano,* while not precisely synonymous, has a meaning strikingly similar to this Greek word, as has also the English word *rutter,* the Portuguese *roteiro* and the French *routier.* The term portolan should not be employed, as has so frequently been done, to designate the charts which especially interest us here; on the contrary, they should be called portolan charts, and this rather than loxodrome or compass charts, as will appear later.

We have no information that the seamen of antiquity were in possession of instruments by which to direct their courses in the open sea. The sun and stars might guide in cloudless weather, but a cloudy sky brought terror to the sailor who had ventured upon a course which led beyond the horizon of known coast lines. Coasting was with the ancient mariners the more common practice, and more useful to them than a seafarer's chart, which might be employed in navigating from port to port across a trackless and unknown sea, would be a written description of the seas over which they were prepared to travel and the coasts they had to visit,—a description of the harbors, the shoals, the currents, the winds, and the facility for anchorage.

Of coastwise navigation in antiquity a few accounts have been preserved to us. The story of Nearchus' voyage from the mouth of the Indus to the Euphrates, in the time of Alexander the Great, is the story of an expedition which was regarded as one of great daring, and worthy the highest praise, but many of the incidents of the expedition show how meagre at that time was the knowledge of real seamanship. The apostle Paul's journey from Cæsarea to Rome was in large part a coastwise journey, and its incidents vividly set forth the dangers and hardships of early navigation. One wonders that so long a time was required for the expedition of the

4

Emperor Justinian to pass from Constantinople to the north coast of Africa, but this expedition, requiring three months, was not directed over the shortest course; instead it too was a coastwise journey, in so far as was possible, leading among the islands of the Ægean, along the coast of Laconia, to Sicily, to Malta, thence across the open sea to Tunis. In each of the expeditions referred to, the periplus must have been the pilot's guide-book rather than the chart.

It is generally accepted that the oldest known periplus is that ascribed to Scylax of Caryanda. Neither the exact year nor the exact century can with certainty be given as the time of its composition. Herodotus relates in Book IV., chapter xliv., of his History, that "the greater part of Asia was explored by Darius, for he wanted to know where the river Indus, the second of all rivers in which crocodiles are found, flows into the sea, and to this end he sent out several trustworthy men, among them Scylax of Caryanda." We cannot, however, be certain that the Scylax here referred to is the author of the periplus. Some of the records, contained in this periplus relate to geographical facts which belong to a time later than that of King Darius, while others in it allude to an earlier day. To all appearances, the greater part of it must have been written shortly before the time of Alexander the Great, and from the standpoint of Macedonia or Greece, seeing, as Kretschmer has noted, the author refers to a road from Corinth on the west coast over the isthmus to "our sea" as forty stadia in length. It includes the entire circuit of the Mediterranean, with a few omissions, beginning at the Pillars of Hercules on the European coast, tracing this coast eastward to the Tanais, thence around Asia Minor and the Levant to Egypt, Libya, and the African coast to a point opposite that of departure, and terminating at the island of Cerne, which island, it is stated, is twelve days' coasting beyond the Pillars of Hercules where the " parts are no longer navigable because of shoals, of mud, and of seaweed." The information given is confined to the immediate coast regions with attention directed to the physical features of the land, to the peoples, the rivers emptying into the sea, to the harbors, headlands, and shoals, with an occasional reference to inland cities in close touch with the coast. The distance from port to port is given, it being stated at the conclusion of his reference to the European coast that one hundred and fifty-three days are needed for a coastwise journey from west to east, and that five hundred stadia might be recorded as a day's sail.

The following quotations will serve to indicate the character of this periplus, which is not a document of great literary worth, though it has a unique value for the history of geography:

" I shall begin," says the author, " at the Pillars of Hercules in Europe, and shall continue to the Pillars of Hercules in Libya, and to the land of the great Ethiopians. The Pillars of Hercules stand opposite to each other, and the distance between them is one day's sail. Not far distant lie two islands by name Gadeira. On one of these is a city which is distant one day's sail. Beyond the Pillars of Hercules which are in Europe, there are many trading stations of the Carthaginians, also mud, and tides, and open seas." He notes that the Iberians are the first peoples to be met with in Europe, and refers to a Greek town which is called " Emporium," adding that " its inhabitants are colonists who came from

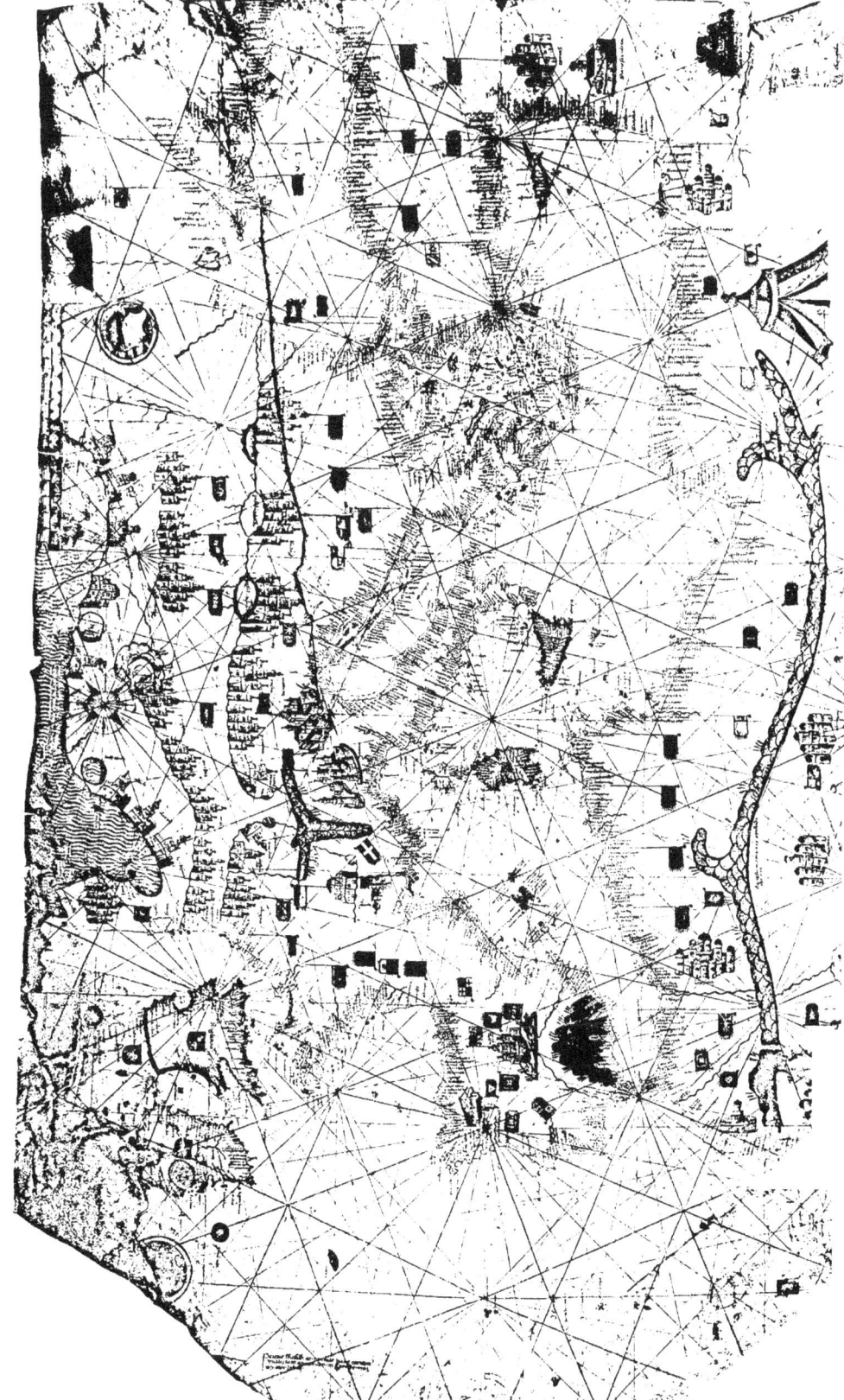

the city of Massilia." "Seven days and seven nights are necessary for coasting along the country of the Iberians." Referring to the Ligurians, it is noted that they are to be found "beyond the river Rhone as far as Antium. Here lies the Greek city and port Massilia, also the colonies of Massilia, Taurnois, Olbia, and Antium. It requires four days and four nights of coasting from the river Rhone to Antium. The entire region from the Pillars of Hercules to Antium is very rich in harbors." Concerning Libya, it is stated that it lies beyond the Conopic mouth of the Nile. "The first peoples to be met with are the Adyrmachidæ. From Thonis the journey to Pharos, which is a desert island, is 150 stadia. In Pharos, there are many harbors, but the ships get drinkable water at Marian. From Pharos to this port is a short sail. Here is also a peninsula and a harbor. To this point is 200 stadia. Beyond lies the Bay of Plinthine. From the mouth of the Bay of Plinthine to Leuce Acte requires a sail of one day and one night, but if you should sail around the head of the bay twice as much time would be required. One next comes to the city of Apis, and as far as this point the country is governed by the Egyptians." In this wise the entire Mediterranean coast region, with minor omissions, is followed with attention directed to the time required in day and night sailing to pass a designated territory, to the inhabitants of the regions passed, to the towns, espe-. cially those of Greek origin, to the geographical features, with an occasional reference to the manners and customs of the peoples. If a chart accompanied the periplus of Scylax, there is left to us no knowledge of it.

In addition to this oldest and most elaborate of all known periploi, certain early descriptions of limited regions have been preserved, as the periplus of the Black Sea by Arrian, who at one time was a prefect of Cappadocia. His description is given in a letter to the Emperor Hadrian. It could hardly have been intended as a pilot's guide-book, though it contains valuable information for those who had occasion to navigate the Black Sea coasts. To the above may be added a fragment by Marcian, probably of the fifth century of the Christian era, which includes a part of the Asia Minor coast, and an anonymous periplus of the Black Sea valuable for its record of distances not only in stadia, but also in Roman miles.

Among those interested in the preparation of charts and sailing directions for seamen, a place of importance is held by Marinus of Tyre. Strangely enough, our knowledge of him and his work is confined to what we may gather from the works of Ptolemy, who lived in the second century of the Christian era. In chapters vi–xx of Ptolemy's geography, Marinus' contributions in this field are critically treated, and from what is there stated, we are justified in inferring that he had carefully examined numerous itineraries and accounts of voyages, that he had prepared a chart to include the regions he described, and that he gave particular attention to the coasts in his work, which was primarily intended for navigators. Ptolemy tells us that in his own work he improved upon that of Marinus, although he gives to the Tyrian full credit for what he had done. We probably have in some of the Ptolemy maps the representations of Marinus. There is reason for believing that there were marine charts passing under the name of Marinus of Tyre, in the second century of the Christian era, which charts were in use by the pilots of the

Mediterranean, the Black, and the Red seas, though such charts seem to have disappeared shortly thereafter.

The Greek periploi were probably employed throughout the Roman period, since in Latin literature no reference is found to original sailing directions for mariners.

A periplus, not second in importance to that of Scylax, and perhaps nearly eight hundred years later, is the so-called Byzantine Stadiasmos. Neither the date when originally written, nor the author is known. It is preserved to us only in part in a manuscript of the tenth century, belonging to the Royal Library of Spain, and once the property of Constantine Laskaris, who, after the middle of the fifteenth century, fled from Constantinople on the coming of the Turk. It has been assigned to the fourth or the fifth century, but internal evidence seems clearly to show that in the form in which it has come down to us there are additions and alterations of later date. The author gives us to understand that it was constructed on the written and the verbal reports of navigators, and that he had set out to present a very exact periplus of " The Great Sea," including a statement of distances from port to port, from island to island, how best to approach them or to direct the course in passing them. It distinguishes between harbors and mere places of anchorage; it indicates whether a port designated is suitable for large or for small vessels, and occasionally states what notice should be taken of the winds in making an approach. Often the details are minute in describing the physical features of certain harbors and coasts, in giving information concerning localities where potable water may be obtained, in pointing out the several important landmarks, such as temples, castles, or other buildings, sand hills, rocks, small islands, headlands, or forests, with an occasional warning that great care should be exercised in navigating certain waters. Apparently it included in its original form the entire Mediterranean and Black Seas. Starting at Alexandria, which city therefore is suggested as the home of the author, it followed the coast to the Pillars of Hercules in Africa, then from the same starting point to eastward, continuing to the Pillars of Hercules in Europe.

It is especially interesting to find that instead of limiting the periplus to a continuous description of the coast of the mainland, a periplus of many of the islands is given, notably of Cyprus and Crete, with which descriptions the Stadiasmos is concluded. Numerous directions are given for sailing from island to island, or from mainland to island, that is, for crossing the sea diagonally; also for sailing in various directions from certain points, as from Rhodes in no less than twenty-five directions, or from Delos in sixteen directions.

Such statements as the last suggest a possible explanation for the introduction of crossing points as they appear later on the portolan charts, though on these charts the radiating points, it is true, have not generally been placed at conspicuous ports, but appear rather to have been inscribed regardless of any particularly important geographical centres.

The Stadiasmos is an exceedingly valuable record for the study of the historical geography of the coast regions covered and may well be considered the most important document known, linking in a sense the older Greek periploi with the later Italian portolans.

A brief extract will serve further to indicate its

character. "1.—Sailing westward from Alexandria to Chersonesus is 70 stadia. Here is a harbor for small vessels. . . . 13.—From Phenicus to Hermæa is 90 stadia; anchor here with the cape on your right. There is water here in a tower. 14.—It is 20 stadia from Hermæa to Leuce Acte; nearby is a low island which is distant two stadia from the land. Boats carrying merchandise can anchor here, entering by the west wind, but near the shore below the promontory there is a wide roadstead for vessels of all kinds. Here is a temple of Apollo, a famous oracle. Near the temple there is water."

In the periplus of Cyprus, which is a part of the Stadiasmos, we find, for example: "297.—Acamas to Paphos, with Cyprus on the left, is 300 stadia. The city is located toward the south. It has three harbors which are accessible with all winds, and a temple of Aphrodite. . . . 304.—From Pedalium to the islands is 80 stadia. Here is a deserted town called Ammochostus; it has a harbor, and may be approached by all winds, but there are low rocks at the entrance. Enter with care!"

In the directions for the circumnavigation of Crete, we find such information as the following: "336.—From Biennon to Phalassarna is 160 stadia. Here is an anchorage, a market-place, and an old city. The island Insagura is distant 60 stadia towards the east. It has a harbor and near the harbor a temple of Apollo. Here is also another island at a distance of 3 stadia, called Mese; it has an anchorage. The third island is called Myle. The channel is deep. It has a market-place."

If to the above periploi of the Mediterranean we add the account of the expedition of Hanno of 465 B. C.

along the coast of Africa, perhaps as far as Sierra Leone, which account contains much information of interest, not unlike in character that given by the periplus of Scylax, and the Ora Maritima of Avienus, describing in like manner the Atlantic coast of West Europe, we have practically all in the way of directions for seamen that is preserved from antiquity.

The middle ages having little or nothing of value to present—a few scattered extracts from earlier writers, a few maps of no special value to navigators,—we may, therefore, pass directly to a word concerning the Italian portolans.

The Italian portolan, as has been stated, resembles the Greek periplus in style and composition. This suggests that these later sailing directions are a development from the former. Such a relationship, however, is not at all easy to establish, since no example is known clearly representing the transition. There is, moreover, in the Italian portolan that which gives it the appearance of a new and an independent production. Very many of the places along the coasts have names other than those in the early periploi; a large number of new names appear; many of the old ones are omitted, which fact suggests that places once known as important had ceased to be so considered; distances are given in miles instead of stadia, and direction is usually recorded.

The number of portolans known antedating 1500 is not large. In all there are about sixteen, some of these being mere fragments, others are very nearly complete for the regions under consideration, and most of them are in manuscript. Those coasts may be said to be included in the Italian portolans which Italian traders were accustomed to visit, that is, the coasts and islands of the

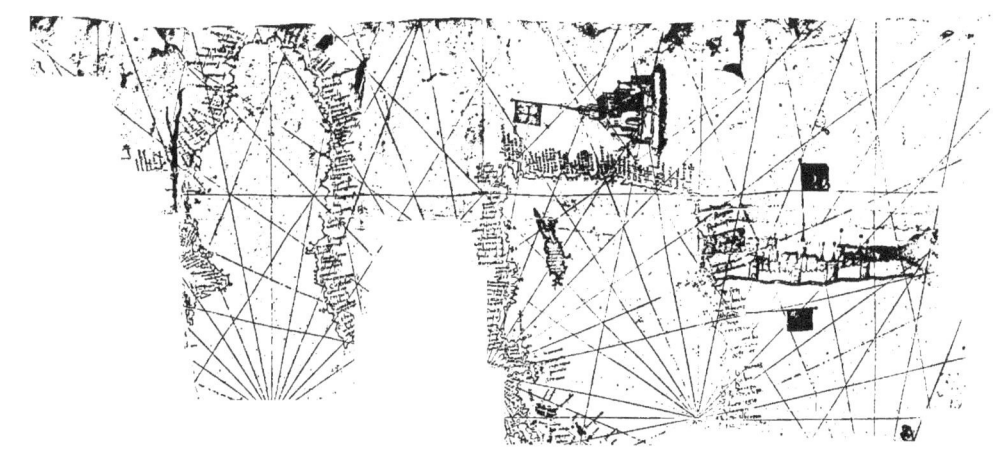

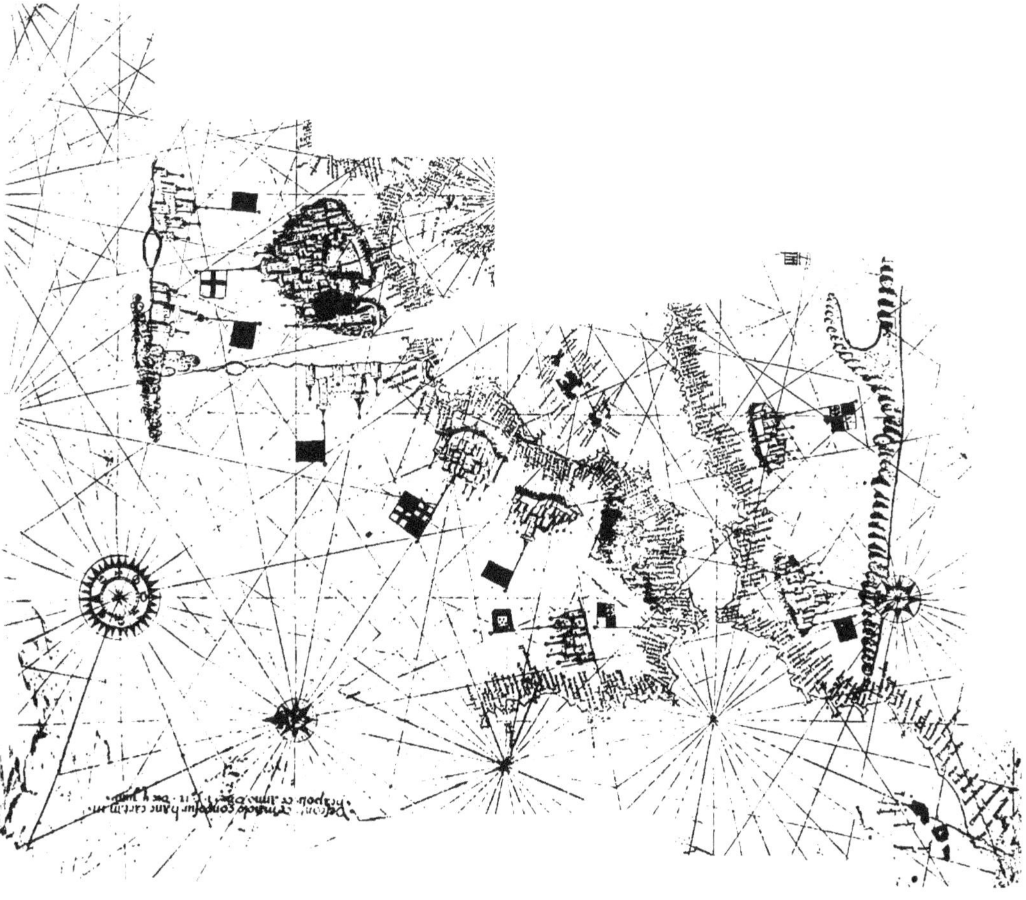

Mediterranean, the Atlantic coast of Europe as far as Flanders, the south coast of England and Ireland, with the Atlantic coast of Africa to the vicinity of Cape Bojador, including the Canary Islands. It is interesting to note that these are the coasts included in the great majority of the portolan charts, with additions, as geographical knowledge expanded, until they became in some instances world charts.

The latest Greek periplus of importance—the Stadiasmos—is of the fourth or fifth century; the oldest of the mediæval portolans is of the eleventh century, and is to be found in the *Ecclesiastical History* of Adam of Bremen, being rather an imperfect sketch of the coast from the mouth of the Maas River to Acre in Palestine. The text of this portolan, together with the text of the others known, may be found in a critical work by Kretschmer, *Die italienischen Portolane des Mittelalters,* pp. 233–552.

The following somewhat free translation of passages contained in the Parma-Magliabecchi portolan of the early fifteenth century will serve to illustrate the character of these Italian harbor books prepared for seamen.

" 45·—From Carminar to Cartagena is 20 miles—northeast by east. Cartagena is a good port at all seasons, before which port there are islands a mile distant. You may pass between any of these islands and the mainland which forms a point. As you enter the port, beware of shoals. Sail close to the middle of the channel, but towards the northeastern shore, where you may anchor. Beware of sailing too close to a shoal recently discovered on the east side. Enter the port, keeping the mainland about two prows' lengths distant, where you have six and six and a half fathoms of water. About the year 1445, it is said, a ship was wrecked here during a calm, though the vessel did not strike a rock. The landmark of Cartagena is a high bald mountain on the east. On the west lies another mountain. Between is the entrance to Cartagena. Near the entrance lies an island, and you may pass between this and the mainland. Passing the island, you enter deep water, and a good anchoring-place."

" 54.—From Sallo to Barcelona is 60 miles east-northeast, quarter east. Barcelona is a city with a shore which lies toward the east having a roadstead with a depth of 22 paces, in front of the city. On the southeast by south of Barcelona is a low place called Lobrigato. In departing, steer to the east from the shore, taking notice of a castle which rises from a depression leading toward Sallo.

The landmark of Barcelona is a high, abrupt, and isolated mountain called Monserrate. When you are northeast of this, continue in that direction, and you will observe a low mountain with a tower on it called Mongich (Montguich). Here is Barcelona."

" 56.—From San Filio to Palamosa is 10 miles east-northeast, quarter east. Palamosa is a good port facing a tower where you may anchor. In case you come from the east, take care of a shoal that is close to the point. From Palamosa to the anchoring-place of Acqua Fredda, 12 miles east-northeast, quarter east. Do not approach nearer the land than one and a half miles by the beacon. The landmark of this bay is a high mountain, bald and cut sheer to the sea, with islands in the distance."

It may be noted that the portolans make their appearance with the awakening of the commercial activities in the coast cities of Europe, notably in the Italian cities, about the tenth or eleventh century, and that for a period of two or three centuries, they served the mariners as a necessary guide in navigation, just as did the periploi in the earlier day. But the quickened commercial activities, coupled with the discovery and use of the compass, were calculated to lead to a speedy sub- . stitution of the chart for the portolan, and portolan charts make their first appearance in what it seems proper to call a very advanced state of development in the years of transition from the thirteenth to the fourteenth century. The stages and the processes of that development we do not know with certainty. We may, however, rest assured that there is a very close relationship between the compass and the portolan chart, as such charts multiply very rapidly in the years following the application of the compass to navigation, but we cannot be quite sure that they owe their origin to the use of the compass. It seems, therefore, not appropriate to call these charts compass charts as has often been done, if thereby we mean to imply that they are based fundamentally on information acquired through the use of the compass. Though the crossing lines may indicate sailing directions, they have not the real character of loxodromes, since they were not constructed on those scientific principles which enter into real loxodrome charts, and furthermore it may well be doubted that the earliest charts of this character were furnished with crossing lines. The term loxodrome chart is likewise not conclusively an appropriate name for them. We may say, in short, that we find in them some of the elements of the simple loxodrome chart, that is, one crossed with lines running from port to port to indicate sailing directions; the elements of a compass chart in which the compass has played a part in determining location and direction; the elements of the ancient periplus—the oldest known pilot-book for navigators; the elements of the mediæval portolan, which is a more elaborate description than is the former of coasts and harbors and sailing directions; and that we find in the portolan the chief corner-stone on which rest the charts here under consideration—hence we may very appropriately call them portolan charts. It may be further stated by way of explanation that *Carta nautica* is the term which is generally employed by Italian scholars in referring to these charts. With them the word *portolano* signifies only a coast or harbor-book. The chart-makers themselves, in referring to their work, most frequently used the word *carta*. On the oldest dated portolan chart, we find the legend " Petrus Vesconte de janua fecit ista carta anno domini MCCCXI," and in a legend on the first chart of his atlas of 1313, we find the word *tabulas* employed. In a chart dated 1605, Maiolo uses the term *carta nauticatoria*. Occasionally the word employed by a chart-maker to refer to his work is merely the personal pronoun, as " Vicentius Prunes in civis Majoricarum me fecit anno 1597."

Portolan charts have been preserved in very large number, of which number near one hundred antedate 1500. In the sixteenth century unaltered in their fundamental character but more highly decorated than those of the fourteenth century, and having additional details, they become far more numerous. With a few exceptions, they are the work of Italian and Catalan chart-makers, a fact which is especially true of the earlier examples. Herein is a most significant witness of the leadership exercised

7. CONTE DE OTHOMANO FREDUCCI, 1537   CHART TWO OF ATLAS.

by the seamen of the Italian and of the eastern Iberian Peninsula; a leadership held for near half a millennium, beginning as early as the eleventh century, and continuing until America had been discovered, Africa had been circumnavigated, and the water route to the Indies had been made known.

In general they are drawn on parchment, as has been stated above, that is, on sheep skin, goat skin, or calf skin, but in time paper came to be used, after which the number of charts of this general character, with additions of numerous details for the interior regions, was greatly increased by means of the printing-press.

They are preserved in two forms, either in single sheets, or in sheets bound together, as an atlas, and these atlases, in a few instances, contain as many as twenty or twenty-five charts. In size the sheets vary from 11 x 15 cm. in the very remarkable charts of the Tammar Loxoro atlas of the fourteenth century to 70 x 148 cm., the size of the large single sheet chart drawn by Pareto in the fifteenth century. The larger world charts, as the Canerio, were drawn on two or more parchment sheets, which were securely joined together. In the case of the single sheet charts, the size, it seems, was most often determined by the size of the skin on which it was drawn, it being true in most cases that the entire skin was used, even the neck being retained, which fact accounts for the peculiar and apparently unnecessary extension of the sheet usually on the left. In the portolan atlases, the several leaves were often made of two sheets or skins pasted together on the rougher surface, leaving the smoother surface for the drawing, which surface received the colors to much better advantage.

These charts, as before stated, include in general the

regions which are referred to in the portolans. The single sheet charts embrace the Mediterranean, and the Atlantic coast of Europe which terminates in the north either at Cape Finisterre or the Scandinavian Peninsula, with a part of the Baltic Sea and the British islands. In the east they include the Black Sea, in the south a part of the Red Sea and the north coast of Africa, with the Atlantic coast of this continent to a point near Cape Bojador.

In the atlases the Mediterranean is usually divided into three sections with one chart for each; one chart includes the Black Sea, and one or two set forth the Atlantic coast regions.

If additional charts were added they usually included a world chart, one or two for the African coasts, one perhaps for the British islands, one for the Baltic, and one or more for the southern Asiatic coasts. A superior example of an enlarged though early portolan atlas is that recently issued by The Hispanic Society of America in facsimile, being a reproduction of a British Museum manuscript, and edited by the author of this monograph.

Portolan charts are projectionless, that is, they do not appear to have been drawn according to mathematical principles or rules, though they were probably based upon measurements and careful calculation. Their striking approach to accuracy, especially for the Mediterranean region, is, as before stated, one of their most remarkable features. No two are alike, and yet they have so many features in common that it appears they are copies of a common original, or that there has been a conscious imitation by each chart-maker as he has set himself to his task of chart-making.

It is well established that most Roman maps were

oriented with the south at the top, an arrangement which is to be met with in the majority of Arabic maps. Maps of the early mediæval centuries have the east at the top, and on the uppermost border a representation of the earthly paradise, as if to give this prominence, it being perhaps the chief factor in determining the orientation. Portolan charts, with rare exception, are oriented with the north at the top, an idea which has since prevailed in all map construction. Herein one seems to find evidence of the influence of the compass in chart construction.

A critical examination will show that in the draughting the chart is turned slightly to the left, the amount being near one point of the compass. As a result of this, geographical localities, on the right of the chart for example, are placed relatively too far to the north. Although there is in this fact the suggestion that the compass had been employed in their construction, or in making the observations on which they are based, and that the declination of the needle had exerted an influence, it may be noted that an acceptable argument has been advanced showing that Constantinople on maps since the time of Ptolemy had been placed too far north by at least two degrees. It appears, therefore, that the error in part is one handed down from an early day. The existence of the error will be readily seen on a critical examination of the location of any selected point in the eastern Mediterranean.[1] As to the length of the Mediterranean from east to west, the near approach to accuracy is also most striking. The error in very many of the sixteenth-century maps, traceable to Ptolemy, and appearing on his maps, is nearly twenty degrees, whereas

[1] *Vid.* Reproduction No. 22 for an exception.

on the portolan charts the error seldom exceeds one degree.

Into a critical consideration of the problems of scale and distance as represented in portolan charts, we shall not be able to enter in this brief description. It is interesting, however, to note in this place that the same scale does not appear to have been employed for the Atlantic coast that was employed for the Mediterranean. Though this fact is not always strikingly prominent, yet it is clearly indicated in a large number of the charts.[1] Herein we may find an explanation for the frequent distortion of the coast regions lying beyond the Straits of Gibraltar, and for the fact that the extension of Europe in latitude is greatly reduced. It may further be noted, as a partial explanation of some of the portolan chartmakers' errors, that it is physically impossible to represent on a plain surface correct distances, retaining at the same time correct latitude and longitude.

A scale of miles divided into fifths or tenths is usually drawn on these charts, often in as many as four or five different places, and frequently on charts of later years in a very elaborate cartouche. It is often very evident that the drafting of such a scale was not done with careful attention to accuracy. Uzielli is of the opinion that it was the Roman mile of 1481 m. which was generally taken as the unit of measurement.

Prior to 1500, degrees of latitude and longitude were seldom if ever indicated on portolan charts, and it may be noted that degrees of latitude are first to be met with on the marine chart of Canerio, recently issued by the author of this paper in size of the original.

A feature of these charts, never failing to attract, is

[1] *Vid.* Reproduction No. 20.

the network of lines with which they are crossed. Though in some instances, a large number of these lines appear to have been drawn as mere fancy directed, it will generally be found that they are arranged according to a carefully devised scheme, and that the lines, usually thirty-two in number, radiate from a number of crossing points, systematically distributed over the chart. The number of crossing points is not always found to be the same, this being frequently determined by the size of the sheet. On portolan charts there will usually be found a central point of radiation about which, in a circle, whose diameter is very nearly the width of the sheet, eight or sixteen other crossing points are represented, each of which is connected with the centre and usually with every other indicated point. On the larger sheets, additional crossing points appear, which points, it will be observed, also fall into a well-arranged system. There was no attempt at a special ornamentation of these crossing points in the earlier charts, but with the passing years, we find now one, now more, especially designed figures for them: wind roses or compass roses these have been called. It is in part due to the peculiar design of these roses that the name compass chart first came into use. While the ornamentation is not always clearly that resembling the compass card, it frequently is such, having that point which is directed to the sidereal or true north extended as if to represent the magnetic needle, but this extension, it will be noted, never indicates the needle's declination. Not until 1532 do we find a printed chart on which the variation of the compass is represented, this being on Ziegler's map of Palestine, and not until 1595 is this declination represented on a marine chart. It is not infrequent that these ornaments are a most striking feature of portolan charts, though adding little to their scientific value.

The suggestion has been made that the crossing lines were originally intended as construction lines, being laid down by the draftsman to guide him in sketching his coasts and in locating his places of special geographical interest, but so few are the instances which might be cited in support of the theory, that one is safe in asserting it to have been the rule with chart-makers to insert the lines after their charts had been drafted.

In the ancient day, it was a common practice with those who had occasion to refer to such matters, to designate each quarter of the heavens by the wind which blew from that quarter. The north was Boreas, the west was Zephyrus, and the number of winds, that is, directions, at first limited to four, was increased in time to eight, then to sixteen. The Italian chart-makers, in general, referred to the winds as eight in number, often representing them on their charts in the wind or compass roses by the first or initial letter of the name. These eight winds were Tramontana, the north, represented by the needle point = Ⓞ , the northeast Greco = G, the east Levante, represented by the Greek cross = ✛, the southeast Scirocco = S, the south Ostro = O, the southwest Libeccio = L, the west Ponente = P, the northwest Maestro = M.[1] We find herein a suggestion that the crossing lines were originally intended to represent the direction of the winds, that is, direction. In time, with the more general use of the compass, the older practice yielded to the newer practice with seamen and direction came to be referred to in terms of compass points rather

[1] *Vid.* Reproduction No. 12 for an excellent illustration.

than in the names of the winds, as for example, *North, N by E, NNE, NE by N.*

The information as to geographical details which is contained in portolan charts, though not extensive, is of much historical interest. It will be observed that the coast lines, in general, have been sketched with care, and usually are continuous, broken only where rivers are represented as emptying into the sea. Bays and headlands, if not accurately inscribed, show that the chartmaker must have had before him information which had been intelligently collected. In some instances, the coast appears as a succession of short curved lines, the result of which is to add a feature of ruggedness. Legends are not inscribed directing attention to rocks and shoals, but these are indicated by small dots or crosses along the coast lines.

Care seems to have been exercised to have all islands represented, and while generally located with a near approach to accuracy, they are often found to be much out of proportion as to size.

The technique of portolan charts is by no means complex, as the geographical information, especially in the earlier charts, is limited to the coast of the mainland, or of the islands. Place names are numerous—for the coast of the Mediterranean alone, the number sometimes exceeds one thousand—and these names, running directly inland from the coast, with rare exceptions were written in small letters, though for the regional names, which were inscribed in the later charts, capitals were employed. Since the names run landward from the coast lines, it will therefore be noticed, as one examines the chart, having the north above, that many of the names are inverted. A large majority of the place names are in black, but it is a striking feature that many are in red, and it is usually the same names so written in the several charts. This fact appears to have no other significance than that a certain special importance then attached, or at least once attached, to the place entered in red.

As these charts were intended primarily for the use of seamen, there was naturally little occasion for attention to the geographical features of the interior regions. These regions, wanting all reference to physical features particularly in the earlier charts, have, therefore, a certain prominence by way of contrast, being blank save for the crossing lines. With the passing years, more and yet more of geographical detail came to have representation on inland regions. River courses in time were represented, though at first with striking inaccuracy: mountain ranges were made to cross certain sections, but clearly attesting the want of exact information: important cities were often made more conspicuous by means of pictures,[1] but cities represented in the interior show a want of knowledge of their exact location. Territorial boundaries do not appear, but many of the separate states bear their respective names, and often in addition are distinguished by an appropriate and highly ornamented coat-of-arms. Castile, for example, has the quartered field with the castle on a red, and a red lion on silver, ground; Aragon, a red standard in a gold field; Portuguese territory, a banner having five dots in a blue field; the Knights of St. John, a silver cross on a red ground; Venice, the gold lion of St. Mark on blue ground; Turkish territory, a banner displaying the half moon; regions remote and unknown, as Tartaria, by a ruler on his throne or an elaborately drawn tent.

[1] *Vid.* For reproduction of picture of Genoa, p. 28, from Bartolomeo Olivo.

In addition to the features just described, legends were often inserted, where space permitted, referring to the products of the region bearing the legend, or to the character of the inhabitants of the same. Much of this information appears to have been derived from Pliny, Solinus, Isidor, or from travellers such as Marco Polo, or Nicolò di Conti. Such legends or descriptive records are, however, generally confined to the world charts of the portolan type which occasionally are to be found in portolan atlases, as for example, in the atlas of Bianco of 1436 or in such as the Catalan world chart of about 1450, belonging to the Royal Estense Library of Modena, Italy. Now and then one finds the earthly paradise represented, as in mediæval cloister maps. Gog and Magog were often located by the chart-makers, as was Prester John, properly adorned as a Christian ruler, and in the Atlantic we frequently find the so-called fabulous islands such as Antillia, Satanaxa, Isla de Man, Brasil, St. Brandan.

Many of the portolan charts are both signed and dated, while many are wanting such inscriptions. Where author and date legend is given it is usually found inserted on the left of the sheet and is very brief, as, "Petrus Roselli composuit hanc cartam in civitate Maioricarum anno domini M cccc lx iij."[1] It is seldom easy to determine the exact age of an undated chart, remembering that such as are dated frequently contain records which clearly indicate carelessness on the part of the chart-maker or the influence of tradition, as may be seen in the representation of a banner, after the authority so indicated in a locality has been overthrown. A noted instance of such false record is the representa-

[1] *Vid.* Reproduction No. 2.

tion of the cross of the Knights of St. John over the island of Rhodes long after that island had fallen into the hands of the Turks. It may further be stated that one is not always justified in giving to a chart a date prior to a known great geographical discovery seeing that such event is not recorded. Portolan chart-makers were generally inclined to make full use in their own records of that which they found at hand. The majority of them were loth to break with tradition or to correct an error, yet we cannot deny to some of them a place of leadership in trans-marine discovery as we find in their charts islands laid down far to the west in the Atlantic, the insertion of which, though not always resting on authentic discovery, unquestionably served to embolden such navigators as were eager for the finding of new lands.

It remains to refer to one of the most attractive features of portolan charts, that is to the colors employed. In some of them the work of the miniaturist of the period is seen at its best. In the earliest examples color was but sparingly used, but with the advancing years it became more and more a feature. The compass or wind rose, at first simple in character, seemed in time to offer to the chart-maker an opportunity to display his sense of the artistic, and not infrequently we find roses which are very elaborate. Banners to be truthful presentations needed color, and they often appear in great numbers and in brilliant tones. Much care was frequently given to drafting designs in which to inscribe the scale of miles, or to the addition of a suitable border for the chart. The effort to emphasize the importance of certain cities led to the addition of fine bits of miniature work to the chart's decorations.

As the crossing lines appear at first to want system or order in arrangement, but on close examination are found to have been laid down in accord with a well-devised scheme, so the color as represented in these lines and in the compass or wind roses seems at first to have been added regardless of rule or of special plan, whereas chart-makers were here most careful in the observance of rule.

It may have been, primarily, for practical reasons that any color scheme was employed at all for the crossing lines. From the multitude of these lines confusion would have prevailed in the attempt to use the chart were the lines of one color. With rare exceptions, it will be found that the lines indicating the eight principal winds or directions are in black, the half winds in green, and the quarter winds in red. For the colors in the several roses or cards, a certain freedom prevailed, especially in the case of those of complicated design. Continental coast lines were generally colored but lightly, though occasionally there was a liberal application of green or blue which was often edged with a line of gold. The coast lines of the larger islands were usually treated as the continental coasts, but smaller islands were entirely covered with red, blue, silver, or gold, and in the case of the smallest of these islands, where numerous, the color was applied so as to produce the most artistic effect regardless of rule. Of the five or six colors employed, red, green, blue, black, gold, yellow, the red seems to be the best preserved.

Seldom was color employed for the larger bodies of water, except in the case of the Red Sea, which invariably exhibits the influence of tradition, being colored red, while on certain world charts of the portolan type, the larger seas and oceans were covered with waving blue or green lines, as may be seen in the Catalan chart of 1450.

Such then in origin, character, and importance are portolan charts with which modern scientific chart or map-making had its beginning. Apparently first constructed in the thirteenth century they multiply rapidly throughout the fourteenth, fifteenth, and sixteenth centuries as before stated, retaining most of the characteristics exhibited in earliest examples. Though remarkable for their near approach to accuracy, it appears not a little surprising that the learned chart-makers of the sixteenth century did not in general accept them at their value until Ptolemy's maps, by actual astronomical measurements, had been shown to be inaccurate. With seamen, however, these manuscript parchment charts remained in favor long after the invention of printing and its use in the multiplication of maps and charts.

BIBLIOGRAPHY

HERODOTUS: History. *Vide* passages relating to Nearchus, Hanno, Scylax, et al.

PLINY: Natural History, iii–vi.

CAROLUS MULLERUS [ED.]: Geographi græci minores,
i., pp. 15–96. *Vide* for the text of Scylax, i., pp.
427–514, for the text of the Stadiasmos.

PTOLEMY: *Vide* various editions of his Geography for
reference to Marinus.

NORDENSKIÖLD: Facsimile Atlas. Stockholm, 1889.
Vide for a review of Ptolemy's contributions in the
field of geography, with references to Marinus.

NORDENSKIÖLD: Periplus. Stockholm, 1897. *Vide* for
a summary of accounts of early maritime expeditions,
with extensive extracts from the periploi, also reference to portolan charts, their character, standard of
measurement, legends, with numerous reproductions.

UZIELLI E AMAT DI S. FILIPPO: Studi biografici e
bibliografici sulla storia della geografia in Italia.
Vol. ii., Roma, 1882. An extensive list of portolan
charts with brief descriptions. Bibliography, pp.
303–312.

FISCHER, TH.: Sammlung mittelalterlicher Welt- und
Seekarten italienischen Ursprungs und aus italienischen Bibliotheken und Archiven. Venedig, 1886.
Contains chapters on portolan charts which are
scholarly.

KRETSCHMER, K.: Die italienischen Portolane des Mittelalters. Berlin, 1909. This contains a summary, not unlike that by Nordenskiöld in his Periplus, but is not in all points in agreement with that work. It contains the texts of the known portolans, with a list of names to be found in portolans and portolan charts. An exceedingly valuable work.

CANALE: Storia de commercio, dei viaggi, delle scoperte e carte nautiche degl' Italiani. Genova, 1866.

WUTTKE, H.: Zur Geschichte der Erdkunde im letzten Drittel des Mittelalters. Die Karten der seefahrenden Völker Südeuropas. Dresden, 1871.

LELEWEL, J.: Géographie du moyen âge, accompagnée d'Atlas et de Cartes dans chaque Volume. 4 parts. Bruxelles, 1852–57.

—— Atlas composé de cinquante planches. Bruxelles, 1850. This published to illustrate the preceding work.

SANTAREM: Atlas composé des Cartes des XIV, XV, XVI, XVII siècles pour la plupart inédites, et devant servir de preuves à l'ouvrage sur la priorité de la découverte de la côte occidentale d'Afrique au delà du cap Bojador par les Portugais. Paris, 1841.

MATKOVICH, P.: Alte handschriftliche Schifferkarten in den Bibliotheken zu Venedig. Wien, 1863.

LUCA, G. DE: Carte nautiche del medio evo designate in Italia. Napoli, 1866.

DESIMONI, C.: Intorno ai Cartografi italiani e di loro lavori manoscritti e specialmente nautici. Roma, 1877.

ERRERA, C.: Atlanti e Carte nautiche dal secolo XIV al XVII conservati nelle biblioteche pubbliche e private di Milano. In Rivista geografica Ital., iii. 1896.

MARCEL, G.: Recueil de Portulans. Réproduction héliographique.

—— Choix de cartes et de mappemondes des XIV, XV siècles. 1896.

Raccolta di Mappamondi e carte nautiche del XIII al XVI secolo [Ongania], Venezia. Seventeen charts are reproduced in photograph.

STEVENSON, E. L.: Marine World Chart of Nicolo de Canerio Januensis. New York, 1908.

More than five hundred portolan charts and atlases are referred to by Uzielli e Amat which are to be found in fifty-four public and private libraries. Referring only to the larger collections, it may be mentioned that ninety-five of these charts and atlases are to be found in Venice, the majority of them belonging to the Biblioteca Marciana, and the Museo Civico; sixty-six are to be found in Florence, chiefly in the Archivo di Stato and the Biblioteca Nazionale; fifty-two are listed as belonging to the British Museum; twenty-six belonging to the Biblioteca Nazionale of Naples; seventeen to the Bibliothèque Nationale of Paris; seventeen to the Archivo del Collegio di Propaganda in Rome; sixteen to the Biblioteca Ambrosiana of Milan. In the other collections designated the number varies from one to fifteen. Comparatively few portolan charts and atlases are to be found in the libraries of the United States. The largest collection is that belonging to The Hispanic Society of America, in which there are thirty-two, the great majority of which are here described for the first time. In the Edward E. Ayer collection of the Newberry Library of Chicago, there are twenty-one; in the Library of Congress three, and in the John Carter Brown Library of Providence there are two remarkably fine atlases.

1. GIACOMO GIROLDI [Probable], early 15th century.

A portolan chart of the early fifteenth century, 53 x 92 cm. in size.   Neither the author nor the exact date can be determined with certainty.   It contains, however, numerous features which suggest that it is the work of the above-named author and its date cannot be far from 1425.   [Fischer, *Sammlung Mittelalterlicher Welt- und Seekarten*, pp. 153–154.]   One other portolan chart by Giacomo is known, dated 1422, which may be found in the National Library of Paris, and three of his atlases bearing dates, respectively, 1426 [reproduced in photograph by Ongania], 1443, and 1446.   This particular chart is here first described.   It includes the Mediterranean, together with the Black Sea, the Atlantic coast of Europe to Friesland, the British Islands, and the coast of Africa from Gibraltar to Cape Cantin.   The chart is crossed by the usual lines, having two complete systems of crossing points, with sixteen points in each group.   The centre of the group on the right is in the Ægean Sea, that on the left near San Sebastian in Spain.   On both the upper and the lower border a scale of miles is given.   The faint indication of a scale on the right, where the tongue extension appears, should not be considered as representing degrees of latitude.

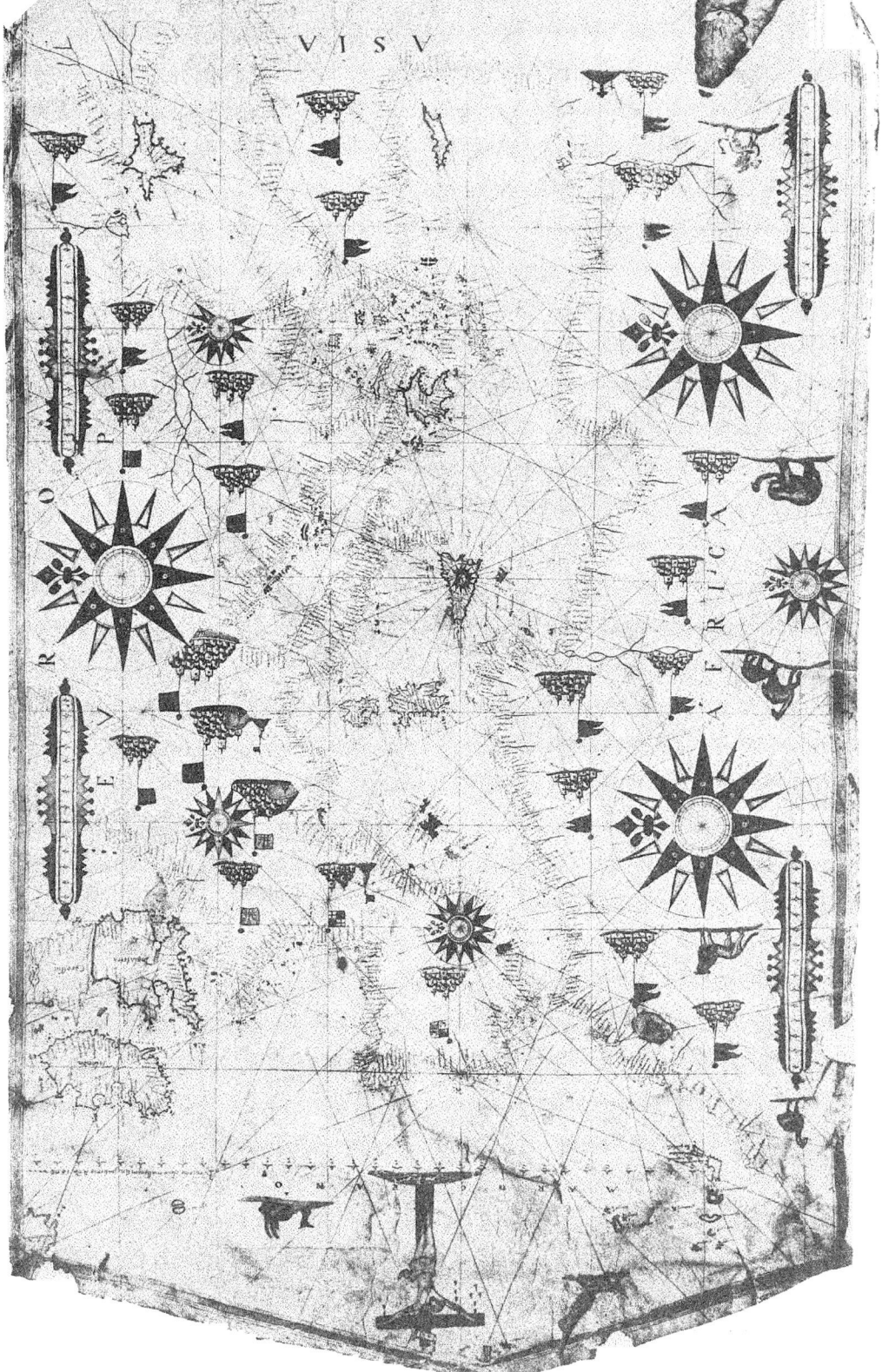

Color has been sparingly used, and compass roses, as well as miniature representation of cities, are wanting. The place names seem to be identical with those given in Nordenskiöld's *Periplus,* pp. 25–44, which names have been taken from Giacomo Giroldi's atlas of 1426. A striking agreement between this chart and the atlas of 1426 may be seen in the physical features of the interior regions which are represented.

In central and eastern Europe the Danube, the Dneister, Dnieper, the Rhine, the Rhone, appear, each of which is represented as having its source in a lake, also the Seine, the Scheldt, the Maas in northern France, the Guadalquivir in Spain, and the Nile in Africa.

With the exception of a few water stains, the chart is well preserved.

2. PETRUS ROSELLI, 1468.

A portolan chart of the year 1468. In size it is 58 x 90 cm. It is a chart of the Catalan type, and exceedingly important on account of its age, and of its geographical and other details. On the upper left of the parchment appears the inscription, " Petrus Roselli composuit hanc cartam in civitate Maioricarum anno domini M cccc lx iij."

Roselli belonged to a famous school of Majorcan cartographers. In addition to this chart, hitherto apparently unknown, four other charts of his may be mentioned, one bearing date 1447, one apparently dated 1462, one dated 1446, and one of the year 1465. The chart has numerous radiating lines, and two small compass roses. A very unusual and interesting feature for a chart of this character is the representation of the winds by four wind-heads.

It includes the Mediterranean, and the Black Sea,

the whole of Europe except Russia and the Scandinavian region, the north coast of Africa, the Canary, the
Madeira, and the Azores Islands. In the Atlantic there
are numerous fabulous islands including " brazil," " illa
de moni " [Man], " antilia," " tamar." These islands
are located as in Bianco's chart of 1436, Pareto's chart
of 1455, and in Benincasas' chart of 1482.

The details which are inscribed in the interior regions
are very numerous. Across the north of Africa stretches
the Atlas range of mountains; the Alps are represented
in Europe; the Carpathian in Austro-Hungary; the
Sierra Nevada in Spain, and Mt. Sinai in northwestern
Arabia. The larger cities are distinguished by groups
of turrets and banners, eleven of which are in Africa;
six are in Europe on a river evidently intended for the
Danube; Venice and Genoa are given their usual prominence, and a conspicuous line of cities is represented
in the Baltic region. The tents in the interior of Africa
give a rather undue prominence to the rulers of that
section. On the north coast of Africa are numerous
Mohammedan banners; on the west coast are those of
Portugal; the Papal banner flies over Avignon; those
of Castile and Aragon give prominence to Spain. Such
decorations as these are particularly numerous on other
parts of the chart. An interesting survival from early
Christian centuries is the idea of giving to the Red Sea
a color appropriate to its name, and this idea, together
with the representation at its northern extremity of the
crossing place of the Israelites, finds expression on most
of the portolan charts which include that region.

Roselli drew much of his information from charts of
the previous century. It is especially interesting to find
that there are numerous features resembling the Catalan

chart of 1375. Though the map is somewhat stained and torn, its colors are well preserved, and its geographical nomenclature, with few exceptions, can be easily read.

3. NICOLAUS DE NICOLO, 1470.

A well-drawn parchment chart of the year 1470, in size 65 x 101 cm. In the tongue of the sheet, which in this instance is on the right, appears the author legend, " Nicolaus de Nicolo M. cccclxx." Nothing appears to be known concerning this chart-maker, who probably was a native of Venice, other than is contained in this one example of his work. In 1882, according to Nordenskiöld, it belonged to Count Pietro Gradenigo.

It includes the east coast of Italy and Sicily, the Adriatic, the Ægean, the Dardanelles, the Sea of Marmora, the Bosphorus, and the southwest coast of the Black Sea, with the west coast of Asia Minor. It is drawn in large scale, has neither compass roses nor scale of miles, though the usual sixteen crossing points with the connecting lines appear. The islands of the Adriatic are made especially prominent, and the lagoons of Venice have been inscribed with great care. None of the cities have been distinguished by picture, nor do regional names appear.

The chart is well preserved, though slightly torn on the edges, and is an excellent specimen of early Italian parchment.

4. ANONYMOUS CHART, late 15th century.

A portolan chart of the fifteenth century, 57 x 91 cm. in size.

Though its author, probably of the city of Venice, is unknown, there is the suggestion in some of its details that it may be the work of Petrus Roselli, as in the representation of the Jordan and other features of that eastern region. Compass lines are numerous, but compass roses are wanting. On the upper and on the lower border a scale of miles is represented. The chart includes the Black Sea, a part of the Red Sea, and the Mediterranean as far west as the Balearic Islands, which region lies in the tongue extension on the left. The coasts are colored, as are also the islands; Rhodes has the cross of the Knights of St. John. Flags and banners are numerous. Some of the important cities, as Venice, Rome, Belgrade on the Danube, Damascus, Jerusalem, and Cairo are represented by interesting colored pictures.

Mt. Sinai is inscribed as an important locality on which is placed the Monastery of St. Catharine. The Danube is laid down as a river flowing directly eastward, having several large islands at its mouth. The Jordan, having its source in a mountain, topped with a castle and a banner, flows through two lakes.

The names are well written, as is usual in red and black, being in Italian and occasionally in the Venetian dialect. It is well preserved, except in parts of its margins.

5. VESCONTE DE MAIOLO, 1512.

A portolan chart of the year 1512, having the very common tongue extension. It is 55 x 90 cm. in size. In the upper corner on the left is a characteristic inscription, " Vesconte de Maiolo composuy hanc cartam in

Neapoli de anno dõi 1512 die 11 jany,'" near which author legend is a well-executed miniature of the Virgin and Child.

Vesconte de Maiolo belonged to a distinguished Italian family of chart-makers whose work has been preserved in numerous examples. Fourteen single charts and atlases are known to have been made by Vesconte from 1504 to 1549. His world chart of 1527, original in the Ambrosiana, Milan, has been issued by this Society in colors of the original, and was included as No. 10 in a facsimile atlas of charts illustrating early discovery and exploration in America, issued by the author of this paper.

This chart of 1512, according to Nordenskiöld, belonged in 1882 to Count Pietro Gradenigo of Venice.

It includes the entire Mediterranean, the Black Sea, the Atlantic coast of Spain from Gibraltar to Cape Finisterre, the coasts of Africa as far south as Cape Bojador, with the Madeira and Canary Islands.

Compass roses are small and are not numerous. On the upper and on the lower border, a scale of miles is drawn. Continental coast lines are colored, as are most of the islands, Rhodes being marked with the cross of the Knights of St. John.

The several countries are designated by appropriate flags. Cities, distinguished by pictures, are numerous, among which Barcelona, Valencia, and Lisbon appear in Spain; Avignon has the papal coat-of-arms, and Genoa south of the Alps, is made especially prominent, being the native place of Maiolo. Six cities are located on the Danube River, which is represented as rising in a mountain in Central Europe, five are in northern Africa, and in Palestine Jerusalem is appropriately represented with an elaborate church edifice topped with a Christian banner. Additional interior physical features are the Atlas Mountains stretching across the north of Africa with striking colors of green and red, and the Sierra Nevada in Spain. The Nile, the Rhone, and the Danube rivers are all distinctly drawn, though not accurately.

The names are in red and black and are in the Italian language.

The chart is well preserved, being injured only in certain parts of the border.

6. CONDE HOCTOMANNO FREDUCCI, 1524.

A portolan chart of the year 1524, being 39 x 60 cm. in size. It is attached to a wooden roller, and has marks of nails in the margin which suggest that it may at one time have been fastened to the walls of a ship's cabin.

On the extreme left is the author legend, "m. vgo. Conde Hoctomanno Freducci de Ancona la fatta nel 1524."

According to Uzielli e Amat, it once belonged to the Marquis Girolamo di Colloredo in Udine. It contains the usual sixteen crossing points with the radiating lines, but is without compass roses.

In the extension of the sheet to the left is given the scale of miles. It has neither degrees of latitude nor longitude indicated.

The chart contains the Black Sea, the northern part of the Red Sea, with the usual representations, and the Mediterranean as far west as the Balearic Islands. Continental coasts are colored, as are most of the islands.

The cross of the Knights of St. John covers the island of Rhodes. Coast names are remarkably well written, and exquisite miniatures of Venice, Damascus, Jerusalem, Mt. Sinai with the Convent of St. Catharine having a subscribed legend, and Cairo adorn the chart. A few rivers are represented emptying into the Black Sea; the Nile delta has been made prominent, and in Palestine the Jordan River and lakes are inscribed, to the east of which is a range of mountains, each peak having a name.

Aside from a few water stains, and slightly torn margins, the chart is in excellent condition.

7. CONTE DE OTHOMANO FREDUCCI, 1537.

An atlas of five portolan charts drawn on parchment, and mounted on pasteboard. Each chart is 35 x 45 cm. in size.

It has an excellent pigskin binding with a title in gold stamped on the front of the cover, " Portolano m. s. 16th century."

The author was a distinguished Italian, and appears to have been a very productive chart-maker, as no less than eight of his atlases are known, and at least five of his single-sheet charts.

On the left of chart four is the legend, " Yhs ma vgo. Conte de Otho maño. Freducci, de Ancona, la fatte nel ano M. cccccxxx7." This atlas, according to Nordenskiöld, first became known in 1882, at which time it was the property of Luigi Arrigoni of Milan.

The sixteen crossing points arranged in a circle about a central crossing, together with the lines which radiate from each point and connect it with every other point save the two on its immediate right and left, give to each chart a very attractive appearance. Each of the five charts has what appears to be a scale of miles marked across each of its four corners, and one compass rose has been drawn on chart three. The borders of the continents are colored, and are represented as a series of large and small curves; islands are red, green, blue, or gold; the nomenclature is in Italian, and written in red or black. Turreted buildings which represent cities are exceedingly numerous, but are small, and are sketched very artistically.

1. Chart one represents a section of the coast of Spain, with the coast of Africa from Gibraltar to a point near the mouth of the Senegal, and the Madeira, Canary, and Azores Islands. Near the parallel of the Canary Islands, and somewhat inland, is a legend of some length. The Azores, the Madeira, and the Canary Islands are practically located on the same meridian. The island of Lancillotto, which is at the centre of the circle of radiating points, has the Genoese cross conspicuously marked.

2. Chart two represents the Atlantic coast of Europe from Gibraltar to Holland, including also in the north England, Scotland, and Ireland, with numerous small islands, among them " Yxola de till," " Montorius," and " isola de man," and to the west of Spain the Azores Islands. England, Scotland, and Ireland are distinguished by regional names artistically drawn. Across Ireland is written a legend as if to explain the significance of a great bay on the west, which is thickly studded with small islands: " Lacus fortunatus ubi sunt insule que dicunt incule sēc beate n̊ ccclxxij."

3. Chart three represents the western Mediterranean with the island of Majorca placed at the centre, and otherwise made conspicuous by its color. Coast names

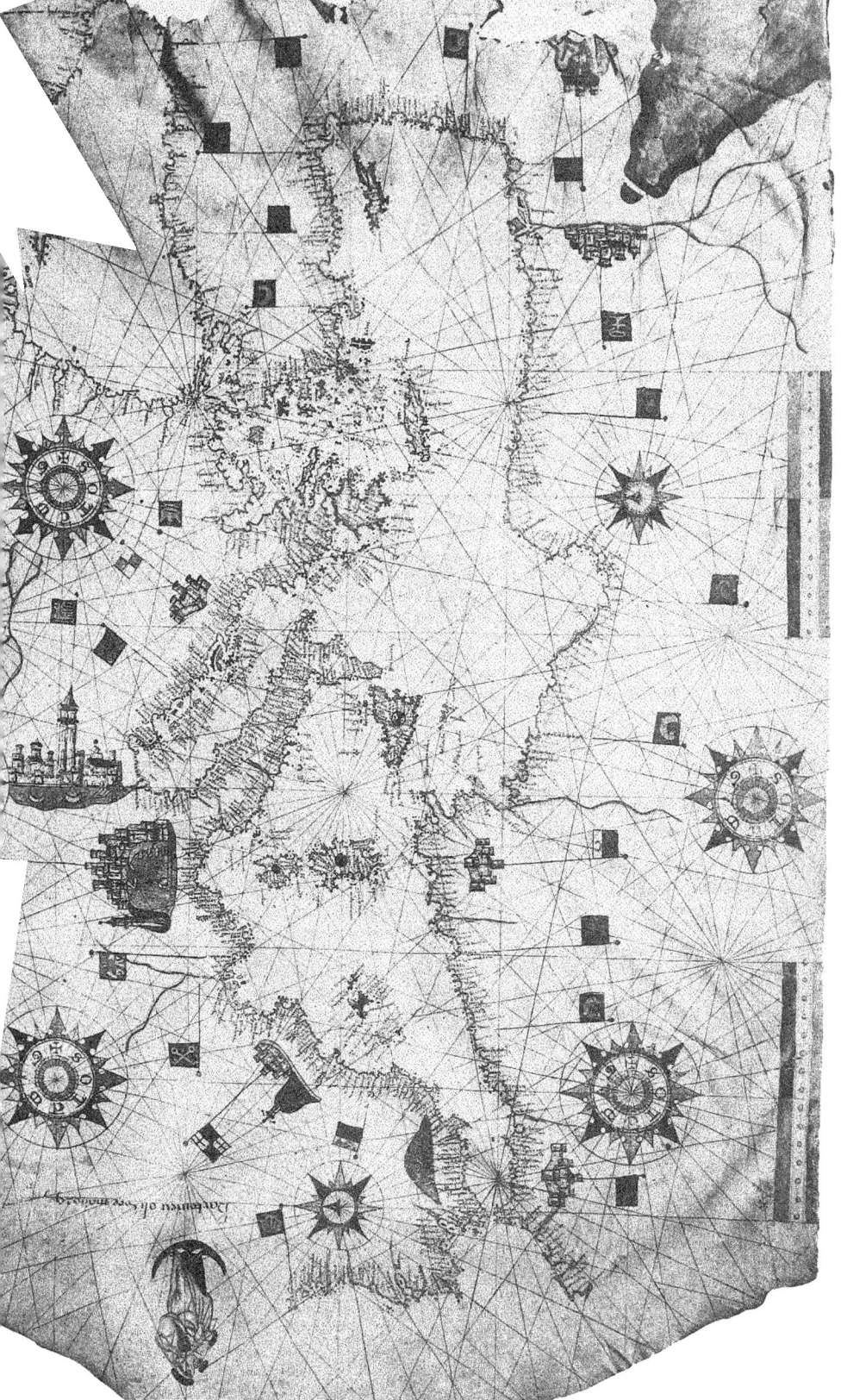

14. BARTOLOMEO OLIVES, 1552

are exceedingly numerous, with an unusual number of names in the Balearic group of islands. The one compass rose of the atlas is to be found on this chart in northern Africa.

4. Chart four includes the Mediterranean from a meridian slightly to the east of Sardinia to the extreme western coast of Asia Minor. The region represented is very striking by reason of its details, a fact especially to be noted of the Ægean Sea. Along the border on the right appears the author legend quoted above.

5. Chart five includes the eastern Mediterranean and the Black Sea, with the bordering regions. Rhodes is conspicuously marked with the cross of the Knights of St. John, and Cairo, though not designated by name, is represented by a large group of buildings stretching along the Nile.

The atlas is remarkably well preserved.

8. ANONYMOUS, first half of 16th century.

A chart of the first half of the sixteenth century, 47 x 74 cm. in size. Though its authorship cannot be determined, it may be referred to as an excellent example of Catalan or Spanish draftsmanship, having its nomenclature in the language of eastern Spain.

While clearly belonging to the sixteenth century, it is based very largely on fifteenth-century originals.

Compass roses are few and are very simple in design.

In addition to the crossing lines running diagonally over the sheet, there are lines which cross at right angles, apparently drawn to represent latitude and longitude, at intervals of eight or ten degrees. These lines, however, are not designated as graduation lines. Four scales of miles are indicated without special ornamentation.

The chart includes the entire Mediterranean, the Black Sea, and a part of the Red Sea, which is colored red, having at its northern extremity the crossing place of the Israelites indicated, the Atlantic coast of the Iberian peninsula southward from Cape Finisterre, and a very limited section of the coast of Africa. Regional names are omitted, but in each of the continents the largest cities are especially distinguished in picture, as Genoa, and Cairo which is called Babelonia. The colored flags and banners inscribed are especially numerous, including the papal banner over Avignon, though not over Rome, the Spanish banner over Spain, Mohammedan banners in Africa and the East, the cross of the Knights of St. John over Rhodes, but not over Malta. Christian powers are represented as holding sway over a part of the Balkan peninsula, Asia Minor, and the coast of the Black Sea.

The Sierra Nevada Mountains in Spain are especially conspicuous, being the only interior physical features represented.

9. ANONYMOUS, early 16th century.

A parchment sheet of the early sixteenth century, 48 x 82 cm. in size. Though its author is unknown, it clearly is of Italian origin, the nomenclature being in the language of the peninsula.

On the left is the tongue extension, apparently used for hanging the chart, on which extension is a miniature picture of Christ on the cross. Graduation is not indicated, but on the upper and on the lower borders a scale of miles is drawn, each in an elaborately ornamented cartouche.

The chart includes the entire Mediterranean, the Black Sea, and the Atlantic coast, beginning at Cape Finisterre, and terminating at a point near eight degrees down the coast of Africa. The coast lines, drawn with a pen, are colored, as are the rivers. As is usual in such charts, most of the names are in black, but many are in red.

Fifteen compass roses of different sizes and designs are drawn, and are connected by the crossing lines. The three continents are each designated by name, which name is written in a scroll. Pictures of cities, flags and banners are wanting, Jerusalem or Palestine being distinguished by the representation of a mountain (Golgotha) with three crosses. Over the islands of Rhodes and Malta the cross of the order of St. John is represented. The chart is well preserved, being torn but slightly on the border, through which nails have been driven, perhaps for attaching it to the walls of a pilot's cabin.

10. ANONYMOUS, early 16th century.

A portolan atlas bound in pasteboard cover, containing three charts of the early sixteenth century. Each chart is 37 x 57 cm. in size.

Neither the author nor the date is known. Judging in particular from the character of the ornamentation, it seems to be a French work [*vid.* No. 22]. The nomenclature is French, Spanish, and Italian. The writing is in rather an unusual running style. Very much of available space is covered with elaborate scroll and feather designs in brilliant colors. Compass roses are numerous, but are not elaborate. Degrees of latitude are indicated on chart one, not on a meridian, but in three sections, with that section on the extreme left indicating degrees from the twenty-second parallel to the thirty-first, the next section near the coast of Spain from the thirty-first to the forty-fourth parallel, and the third section some distance to the west of the coast of Ireland from parallel forty-four to fifty-eight.

1. Chart one includes the Atlantic coast regions from Holland—" Olanda "—to a point near Cape Verde, with the British Islands, the Azores, the Madeira, and the Canary Islands, and the Mediterranean to the Gulf of Genoa.

The regional names include " Europa," " Olanda," " Irland," " Iscotia," " Inglaterr," " France," " Spagnia," " Africa," " Barbaria." In the Atlantic to the west of Spain is the legend, " L'ocean occidental," and near it an elaborate ornament containing a shield with the French lily topped with a royal crown. In Spain is the Spanish shield with the imperial double eagle. Much attention was given to the ornamentation of the sheet.

2. Chart two includes the Ægean with the bordering coasts. The names " Natolia," " Candia," " Morea," " Grecia," " Romania " are inscribed. The sheet is somewhat injured by water stain.

3. Chart three includes the entire Mediterranean, and, like the preceding, is over-decorated. " Europa," " Asia," " Africa," " Barbaria " are inscribed. The chart is much faded.

The atlas is not one of great value; though containing practically all that may be found in the better Italian portolan charts of the period, it was apparently designed for display rather than for use.

11. BAPTISTA AGNESE, early 16th century.

A portolan atlas of the first half of the sixteenth century, containing fourteen charts preceded by representations of coats-of-arms and astronomical tables. Each page is 28 x 41 cm. in size.

The atlas is neither signed nor dated, but the workmanship is so strikingly characteristic of Baptista Agnese that one can hardly be in error in ascribing it to him. Its date cannot be far from 1545. Agnese was a prolific chart-maker, many of whose atlases are extant. He exhibited remarkable skill as a draftsman and miniaturist, and held a foremost place among the Italian chart-makers of his time. His work, however, appears to have been done rather for the libraries of princes than for the use of mariners.

The atlas is one of his largest, containing not only those charts which usually are to be found in his atlases, but certain important additions as Nos. thirteen and fourteen.

The several charts have retained their brilliant colors, with blue coast outlines, and with the numerous small islands in red, blue, or gold. Interior regions on some of the charts contain numerous vignettes and figures, with far more than the usual references to geographical features, though having withal numerous and curious inaccuracies.

Page two of the atlas contains two coats-of-arms similar in general design, with a third on page three of like character. Declination tables are given on page four, an armillary sphere is artistically sketched on page five, and on pages six and seven the circle of the zodiac with the several signs very artistically designed.

1. Chart one, occupying double pages, as do all the charts, being in size within the narrow black border line 35 x 50 cm., includes the Pacific Ocean, with America on the right, and on the left a small section of the coast of Asia, with the " insule Maluche." The Atlantic coast of America is represented from Labrador to the Strait of Magellan, with an omission on the extreme east of Brazil. The Pacific coast includes Lower California and the Gulf of California which is colored red, thence it extends southward to about latitude 12°, omitting the remaining section of the coast to the Strait of Magellan. Degrees of latitude and longitude are indicated, as are also the tropics and the equator. In the centre of the chart is a combined compass and wind rose encircled with the sixteen crossing points from which the usual thirty-two lines radiate.

2. Chart two includes the Atlantic Ocean with Africa, a part of Asia and of Europe on the right, and on the left the east coast of North America, the east and west coast of South America, omitting the Pacific coast from latitude 12° to the Strait of Magellan. The nomenclature is very rich, both for the Old and for the New World, and is remarkably well preserved. The chart contains the crossing lines, degrees of latitude and longitude as in chart one.

3. Chart three includes the Indian Ocean, with the African coast from the Gulf of Guinea eastward, and the southern Asiatic coast to China. The general features of the chart are the same as are represented in the preceding.

4. Chart four includes the greater part of the continent of Europe with Scandinavia on the northern border, the Baltic, the British Islands, the coast of France, the north coast of Spain, northern Italy, and the Dalmatian coast. There are numerous fine miniature representations of kings on their thrones. Numerous rivers are represented in heavy blue lines, though with little care for accuracy, and the mountains of southern Europe from the Pyrenees to the Balkan peninsula are inscribed. Numerous regional names appear. The features of chart one are included, omitting indications of latitude and longitude.

5. Chart five includes the Spanish peninsula, with northwest Africa, the Balearic, the Madeira, and the Canary Islands. Numerous cities in picture, and rivers in heavy blue lines are represented in Spain. The Atlas range of mountains is represented in northern Africa. The chart has the usual crossing lines, but no indication of degrees of latitude or longitude.

6. Chart six represents the Mediterranean region from the west coast of Greece to Gibraltar. It contains the usual portolan chart features. It is slightly water-stained, but all names can be read with ease.

7. Chart seven represents the eastern Mediterranean from Sicily to Palestine, with the south coast of the Black Sea and the northern section of the Red Sea, which is colored red. The island of Rhodes still retains the cross of the Knights of St. John.

8. Chart eight represents the Black Sea with the coast line in blue, except that of the Crimea, which is colored green. The chart is remarkably clean and attractive.

9. Chart nine represents Italy with the Adriatic

coast of Dalmatia. This chart differs greatly from the preceding, the author attempting to produce not only a sea chart but a land map as well. The entire surface is covered with a pale-yellow color shading to light green. Mountain ranges are especially conspicuous. The Po River with its numerous branches fills the low plain of northern Italy. The Tiber and the Arno rivers have a source in the same lake. Sicily, Sardinia, and Corsica are very prominent.

10. Chart ten represents the Ægean Sea as it appears in most portolan atlases. The island of Rhodes is conspicuous with the red ground and the gold cross. Other large islands have the entire surface either green, red, or gold. Crete has a city with harbor prominently marked on the north coast.

11. Chart eleven is a world chart including practically the same as may be found in Nos. one, two, and three. The whole of the north of Europe is, however, represented with the Scandinavian peninsula stretching off to northward. There are five compass or wind roses, and the usual arrangement of crossing lines. The tropics and the equator are represented, but graduation is wanting.

12. Chart twelve is a characteristic Agnese world chart, oblong in shape, having the equatorial diameter nearly twice that of the polar. Fifteen parallels are represented and twenty-four curved meridians. It is more or less a conventional world chart, since the author clearly made little effort to be strictly accurate. Numerous regional names are inscribed, and a few interior geographical features appear, as mountains in Asia, "Mons luna" in Africa, the Pyrenees in Spain, and the Andes in northern South America. There are also

numerous rivers, as the Nile, the Indus, the Ganges, the Volga, the Danube, and, in the New World, the Amazon and the La Plata. Magellan's route is indicated, as is also the route from Spain to Peru, crossing the New World at Panama. Twelve artistic wind heads are arranged about the chart, each wind having its appropriate name. Compass roses as well as crossing lines are wanting.

13. Chart thirteen represents Palestine. Like Nos. nine and fourteen, it is both a marine and a land chart. It has a ground color of very light yellow shading to green, though the color has not been evenly applied. Mountain ranges and isolated mountain peaks are very numerous, and are made especially conspicuous. The several towns and villages are each represented by an artistic vignette, and Jerusalem appears as a group of buildings surrounded with a wall. The Jordan River and the lakes are colored blue, and are much magnified in size. As drawn, it places the east at the top of the sheet, though a compass rose is inscribed, which indicates the north.

14. Chart fourteen includes the Scandinavian peninsula with the Atlantic islands to the northwest and the west, and with the Baltic Sea and its neighboring regions to south and east. The chart has the same ground color as Nos. nine and thirteen. There are numerous artistic representations of rulers on their thrones. Two well-drawn ships sail the Atlantic waters, and a sea monster appears off the Norway coast. Very small and artistically-drawn buildings representing cities and towns are numerous. One wind head is represented at the north, but all crossing lines are omitted. The chart is slightly water-stained at the top.

12. BARTOLOMEO OLIVO, after 1550.

This portolan chart of the sixteenth century is 86 cm. in length by 51 cm. in width.

Its author was a member of the famous Oliva family of Majorca. On this chart his name is inscribed in the upper left corner, " Olivo mallorqin En. Palermo Año 1520." The last three figures are inscribed over an erasure, and only the figure 1 is the original. We find here one of the numerous attempts at date forgery. Sometimes for a specific reason, generally to give a ficti- tious value to a chart, a date is found altered to one earlier, often to one later. While the argument is not conclusive, it appears to have been drawn after 1526, as the cross of the Order of St. John appears on the island of Malta, and perhaps near 1581, as the Spanish flag alone appears on the Iberian peninsula, whereas on Domingo Olives' chart of 1568 Lisbon is likewise so distinguished, by a Portuguese banner.

Compass roses are numerous, and elaborate, and it is interesting to observe that the central crossing is in Sicily. On the western border is a representation of Christ on the cross. The scale of miles is indicated twice at the top, and twice at the bottom of the chart. Here we also have it distinctly indicated that the same scale was not used for the Atlantic coast that was used for that of the Mediterranean. Degrees of latitude are marked on a meridian passing about five degrees west of the coast of Spain.

The chart includes the Mediterranean, the Red and the Black Seas, the Atlantic coast region as far north as Holland, with England, Scotland, and Ireland, and the African coast to Cape Bojador. The author has not

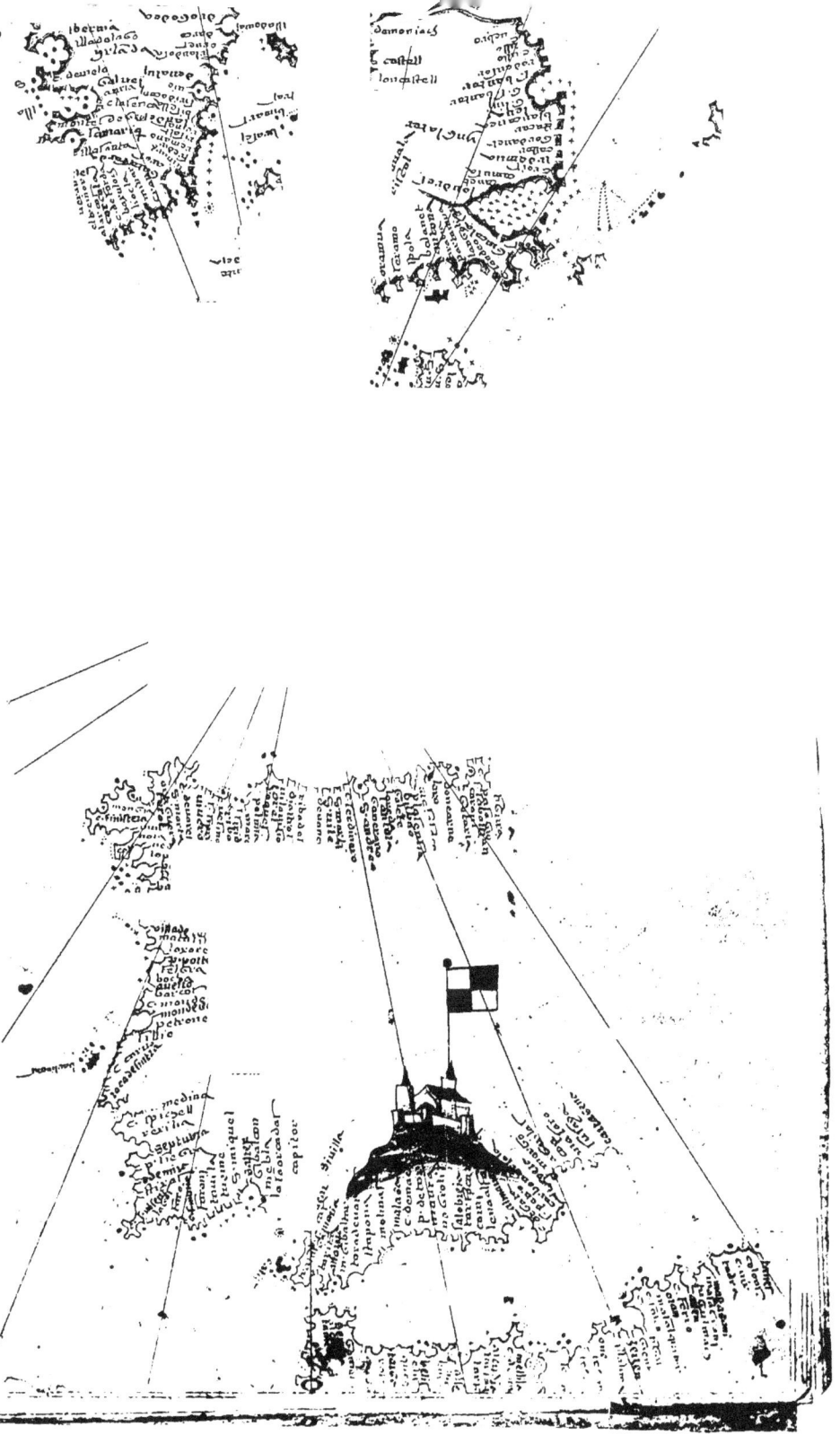

16 JAUME OLIVES, 1563.   CHART THREE OF ATLAS

given a colored border to his continental coasts, but has added color liberally to his islands. The three continents are distinguished by names; cities especially distinguished by picture are numerous both along the coasts and inland; numerous river courses are shown, though not from accurate information. An ocean monster appears to the southwest of Ireland, and in Africa, the ostrich, the lion, the camel, and the elephant are well represented.

The chart is well preserved, being only slightly torn on its borders.

13. HIERONYMO GIRIVA, after 1550.

A portolan chart with pasteboard cover of the second half of the sixteenth century. It is **33** x **65** cm. in size. It is neither signed nor dated, but internal evidence suggests that it is the work of Giriva. On the interior of the front cover has been pasted the engraved book-plate of Conde de Vislahermosa. The chart includes the Mediterranean with a very small section of the Black Sea coast, and the Atlantic coast from " lisbona " to Cape Cantin.

Compass roses are numerous, being inscribed at most of the sixteen crossing points. The central rose is placed in the island of Sicily. In each of the four corners a scale of miles is drawn in a waving scroll ornament. In the upper corner on the left is a representation of Christ on the cross. Eleven cities are distinguished by miniature pictures, including " lisbona " and " barselona " in Spain. Over each appears an appropriate banner. Mountains are represented in northern Africa, and Golgotha with the three crosses. A few of the important rivers are represented in their lower courses, as the Nile,

the Rhone, the Guadalquivir. The chart is well **pre-**
served, but the decorative work is not that of a first-class
miniaturist.

14. BARTOLOMEO OLIVES, 1552.

A portolan chart of the year 1552, in size 49 x 75 cm.

On the left is the author legend, " Bartolomeo olives
maijorq; 1552," and in the tongue extension a miniature
of the Virgin and Child.

Bartolomeo was one of the most distinguished members of the Majorcan family of Oliva, which family has
a place of prominence among early chart-makers. A
large number of his charts are recorded by Uzielli e Amat,
and by Nordenskiöld. Compass or wind roses are
numerous at the crossing points, five of which are large
and beautifully executed. In each of these the initial
letters for the eight winds appear, beginning at the north
with the needle point ◉ thence to the right about the
circle G = NE; ✠ = E; S = SE; O = S; L =
SW; P = W; M = NW. Twice on the upper and
twice on the lower border the scale of miles is inscribed.

The chart includes the Mediterranean, the Black Sea, a
section of the Red Sea with the crossing-place of the Israelites indicated, and the Atlantic coast from Cape Finisterre
to Cape Nun. Seven cities are represented in picture,
Venice, Genoa, and Cartagena being especially prominent,
and the banners are very numerous. Interior physical
features inscribed include the Sierra Nevada Mountains,
very conspicuous in southern Spain, Mount Sinai topped
with the Convent of St. Catharine, also the Nile, the
Danube, and the Rhine rivers, with the numerous others
distinctly indicated in their lower courses.

The chart is remarkably well preserved, having only its margin on the right slightly torn.

15. GIOVANNI MARTINES, after 1560.

A portolan atlas of seven charts, 18 x 25 cm. in size, bound in dark leather. Each chart, however, occupying two pages, is 24 x 34 cm., including a plain narrow red border. Across the upper part of chart one is the author legend, " Joan Martines en Messina añy 1562," the figure 2 being written in after an erasure. It is hardly probable the date as it appears is correct, though it doubtless is one of his earlier atlases, the oldest hitherto described being dated 1567. He was one of the foremost chartmakers of his day, there being extant a considerable number of single charts and atlases bearing his name. On the back of the front cover are the words in gold " Carta Navigatoria," and on both front and back the outline design for a coat-of-arms, with the letters M. P. G., the initial letters of a former owner, " Michele Petrocochino del quondam Georgio." Fifteen pages of manuscript in Italian and written in a bold hand, relating to astronomy and geography, have been bound in with the charts.

The several charts have the usual characteristics of the type. Coast lines are in gold, or in blue and gold. A few minor geographical features are represented, as regional names, and the lower courses of rivers. Each chart, except the first, has a scale of miles, and at least one compass rose.

1. Chart one represents the world in two hemispheres, each having an equatorial diameter of 167 mm. and a polar diameter of 165 mm. Meridians and paral-

lels are drawn at intervals of fifteen degrees. Coast lines are in gold, and numerous original names appear in both the Old and the New World. The Strait of Anian (Bering) is represented, which name appears in northeast Asia. The great austral continent "terra incognita" is sketched in outline. It is a chart to which much interest attaches.

2. Chart two includes the southern coast of the Spanish peninsula, the west coast of Africa to about latitude 15° north, with the Madeira and the Canary Islands. On this and succeeding charts three crossing points and three only are represented, from which thirty-two lines radiate. One of these points is located at the centre, one at the right of this, and one at the left, but all are on the same parallel or line crossing the chart from east to west.

3. Chart three includes the northwest coast of Europe, with England, Scotland, and Ireland well represented. The names are numerous and well written.

4. Chart four includes the south coast of Ireland and England with the coast of Holland, France, and northern Spain.

5. Chart five includes the Mediterranean from the eastern coast of Spain to the west coast of Greece, also the Balearic Islands, Corsica, Sardinia, and Sicily, with a small section of the extreme north coast of Africa. On this chart the names are particularly numerous.

6. Chart six includes the Ægean and the eastern Mediterranean.

7. Chart seven includes the Black Sea drawn in large scale.

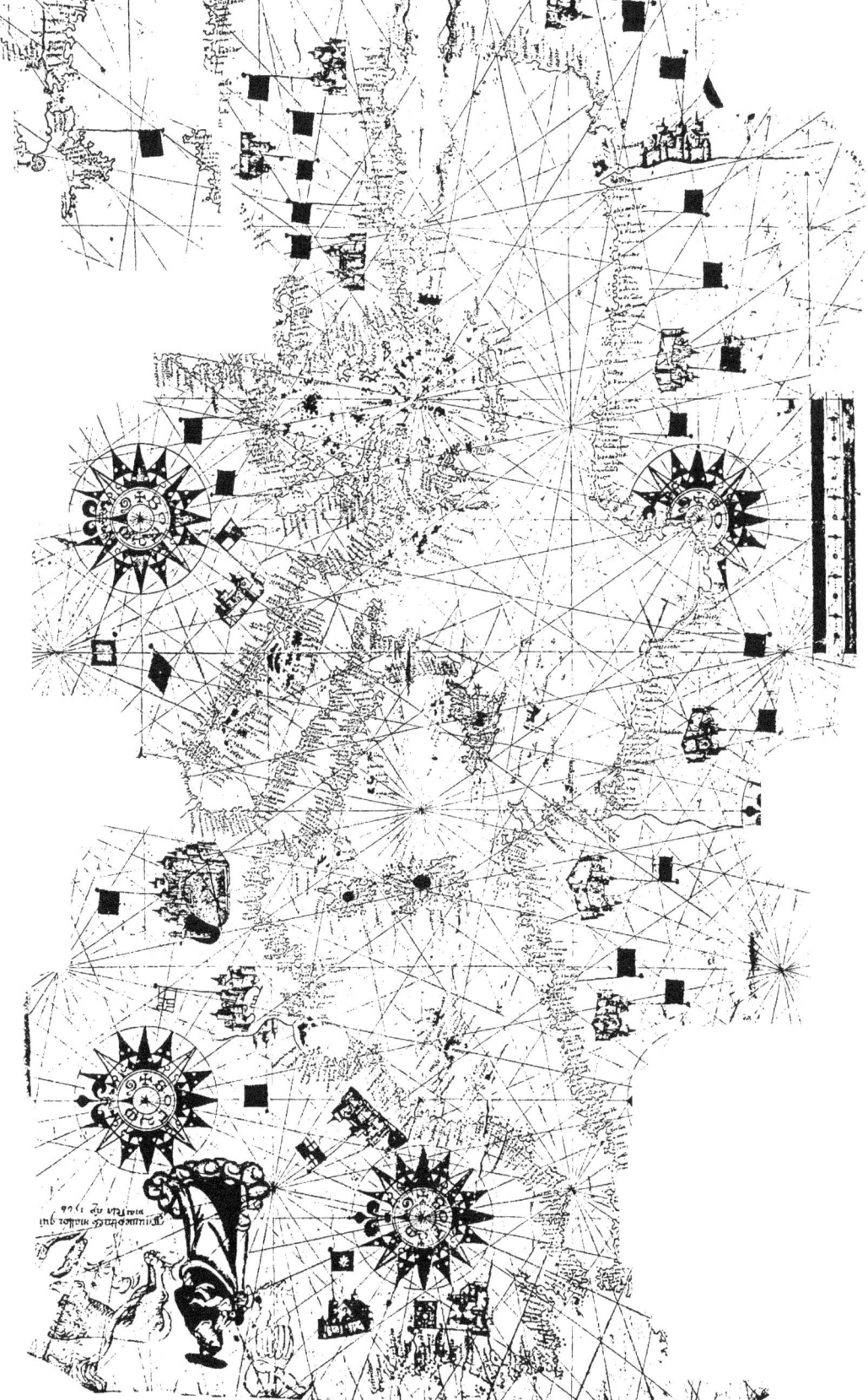

This atlas of six charts dated 1563 is 19 x 23 cm. in size, though each chart, occupying double pages, measures 23 x 36 cm.

It has an excellent leather binding, with the entire front and back of the cover very artistically decorated with contemporary tooling.

It is a characteristic bit of work of a very distinguished member of the Majorcan family Oliva, which family had numerous representatives in the ranks of the early chart-makers.   On the last double page appears the inscription, " Jaume oliues mallorchi en napoli any 1563."

All names have been inscribed with great care, partly in Italian and partly in Catalan.   The usual portolan chart colors have been employed all of which are well preserved.   From a central point on each chart thirty-two lines radiate with one exception, all other lines having been omitted.   It is interesting to find that on the last double page a very large compass rose has been drawn filling almost the entire sheet, and on the first double page a circle has been drawn with the radiating lines, suggesting that the author had intended these as construction lines for a chart which had never been drawn.   Coast lines, in most instances have been colored, to which has been added a gilt border.   There are numerous miniature representations of cities and banners on each of the charts.   With the exception of sheet five, a part of which has been cut away, the atlas is remarkably well preserved.

1.   Chart one represents the eastern Mediterranean, omitting the Levantine coast.   Chios and Rhodes have the Christian cross while all the banners represented are the Mohammedan.   The chart has the thirty-two radiating

lines and instead of a wind or compass rose it has the usual initial letters for the eight principal winds properly placed near the border of the sheet, which arrangement is very unusual.

2. Chart two includes the Atlantic coast from " c. finisterr " to " c. blancho " with the Canary and the Madeira Islands.

3. Chart three includes the western coast of Europe from " cartagena " to " dascie " on the North Sea coast, with a small section of the northwest coast of Africa, also England, Scotland which is separated from the former by a strait, and Ireland. There is a central compass rose from which thirty-two lines radiate. Three cities are especially distinguished in the Iberian peninsula, each with picture and banner.

4. Chart four represents the western Mediterranean from Sicily to the Strait of Gibraltar. The Spanish, Papal, and Mohammedan banners are prominent features of the chart, and the city of Genoa is made especially conspicuous in picture.

5. Chart five represents the middle Mediterranean region, including the Adriatic and the Ægean Seas. This chart has been somewhat mutilated on the upper or northern section.

6. Chart six represents the Black Sea and the extreme eastern Mediterranean coast. In the centre is a compass rose from which thirty lines only radiate. Banners are numerous, most of which are Mohammedan.

17. JAUME OLIVES, 1566.

A portolan chart of the year 1566, rectangular in shape, and in size 46 x 69 cm. On the upper left appears the Madonna and Child resting in the clouds, near this picture a lion, which is represented as tearing an animal in pieces, underneath which is the inscription, " Jaume Olives, Mallorquien Marsela ãy 1566." Jaume Olives was a member of the famous family of Oliva which first came into prominence in the island of Majorca. Other distinguished members of this family were Bartolomeo and Domingo Oliva, each being the author of numerous portolan charts. Five other charts of this author are known, which represent the Mediterranean, one of which, bearing date 1557, is in the University Library of Pavia, one dated 1559 is in the National Library of Naples, one of 1561 is in the Vittorio Emanuele Library in Rome, one of 1563 is in the Museo Civico of Venice, and one of 1566 in Marseilles. The one here described appears to have been his last. Seven large compass roses are included in the circle of sixteen crossing points. Neither latitude nor longitude is indicated. Four scales of miles are drawn.

The chart includes the entire Mediterranean, the Black Sea, the Red Sea with the indicated course of the Israelites at the northern extremity, the Atlantic coast of Spain from Cape Finisterre, and the coast of Africa to Cape Blanco.

Important cities are made prominent, notably Genoa and Venice, and brilliantly-colored flags and banners are very numerous.

The names are written in very small letters, and are numerous. The corners on the left of the sheet have been torn, but not to such extent as to injure the contents of the chart. It is one of the most valuable of the collection.

18. GIOVANNI MARTINES, 1582.

A portolan atlas of the year 1582, containing five charts each 32 x 48 cm. in size, bound in pasteboard cover.

On chart four appears the inscription, " Joan Martines en Messina Añy 1582."

A number of charts and atlases by this author are known, all of which are exceedingly well done. In the front of this atlas is pasted a brief description of the several charts by E. F. Jomard, editor of the famous atlas " Monuments de la Géographie." There are three or more compass roses on each chart, some of which are elaborately executed. The usual intricate crossing lines are inscribed, and on each chart a scale of miles drawn on a great waving scroll. The nomenclature is especially rich.

1. Chart one includes the eastern Mediterranean, the Ægean, and the Black Seas. Over Jerusalem waves a flag with the cross, over the Crimea the Genoese flag, which flag also appears prominent at the entrance to the Black Sea. The Red Sea is a conspicuous feature, and Cairo appears as a many-turreted city on the Nile.

2. Chart two includes the central and western Mediterranean. It contains all the characteristic features of the preceding, with Venice, Genoa, Marseilles, and two cities in northern Africa, made prominent by groups of turreted buildings.

3. Chart three represents the coast region of western Europe from Gibraltar to Denmark and Iceland. This chart is especially interesting. England and Scotland, as on contemporaneous charts, are represented as separated by a strait. Iceland appears on the extreme

northern border, southwest of which is "Frixlandia," with a few other names to be found on the Zeno map of ·1558. "Isla de Brasil," with its usual peculiar features, is located to the southwest of Iceland. The Iberian peninsula is especially distinguished by its turreted cities and conspicuous banners. About four degrees to the west of Spain the prime meridian is drawn across the chart, on which degrees of latitude are very distinctly indicated.

4. Chart four presents the west coast of Spain and the coast of Africa from Gibraltar to the mouth of the Senegal, which point is conspicuously marked with a Portuguese banner, as is also the city of Lisbon. The prime meridian on which the degrees of latitude are marked is represented much to the west of the Canary Islands, and is at least ten degrees farther west than on the preceding chart. It is on this chart that the name of Martines appears.

5. Chart five represents the west coast of Africa from the mouth of the Senegal to the Cape of Good Hope, the last-named point being especially marked in large capitals "CAPO DI BONA SPIRANZA." Numerous Portuguese banners are inscribed along the coast. The prime meridian, if such it is intended to be, runs slightly to the west of Africa, north of the equator, while south of the same it is represented as starting at a point at least four degrees farther to the east.

The atlas is well preserved in all its rich details.

19. ANONYMOUS ATLAS, late 16th century.

A French atlas of portolan charts of the second half of the sixteenth century in brown leather cover. It contains four charts each occupying double pages 49 x 61 cm. in size. On the outside of the front cover is stamped "Ex libris Luigi Arrigoni Mediolani." Its author is unknown, but it corresponds in all important particulars to the contemporaneous work of Italian chart-makers.

Compass roses are numerous and some are very large. Certain designs for banners, as well as the cartouches in which the scale of miles is inscribed are brilliantly colored and elaborately executed. The coast lines are colored, certain parts of which are unusually heavily marked.

The coast nomenclature is very full, and regional names, especially, are in French. French portolan charts of the period are not numerous, and this work is one of the most valuable of the kind known.

1. Chart one drawn on a large scale includes the western Mediterranean, the Atlantic coast of Africa to Cape Cantin. Many regional names are inscribed as "Europe," "Spagne," "Genes," "Provence," "Catalogne," "Valence," "Granade," "Andalvzie," "P:Gal," "Afrique," "Tunis," "Arger," "Barbarie," "Fex."

2. Chart two includes the middle Mediterranean and the Adriatic. The coast of France, of Tunis, of the island of Sicily, and the opposite mainland of Italy, the Adriatic coast of Tuscany, the coast of Istria, and the coast of western Greece, all are made especially prominent by heavy coloring.

Regional names are numerous, including "Europe," "Genes," "Venise," "Italie," "Tuscane," "Istrie," "Dalmatie," "Tunis," "Tripoli." "Greece," and "Afrique" are inscribed in a conspicuous cartouche. As inserted in the atlas, the north is at the right.

3. Chart three includes the eastern Mediterranean,

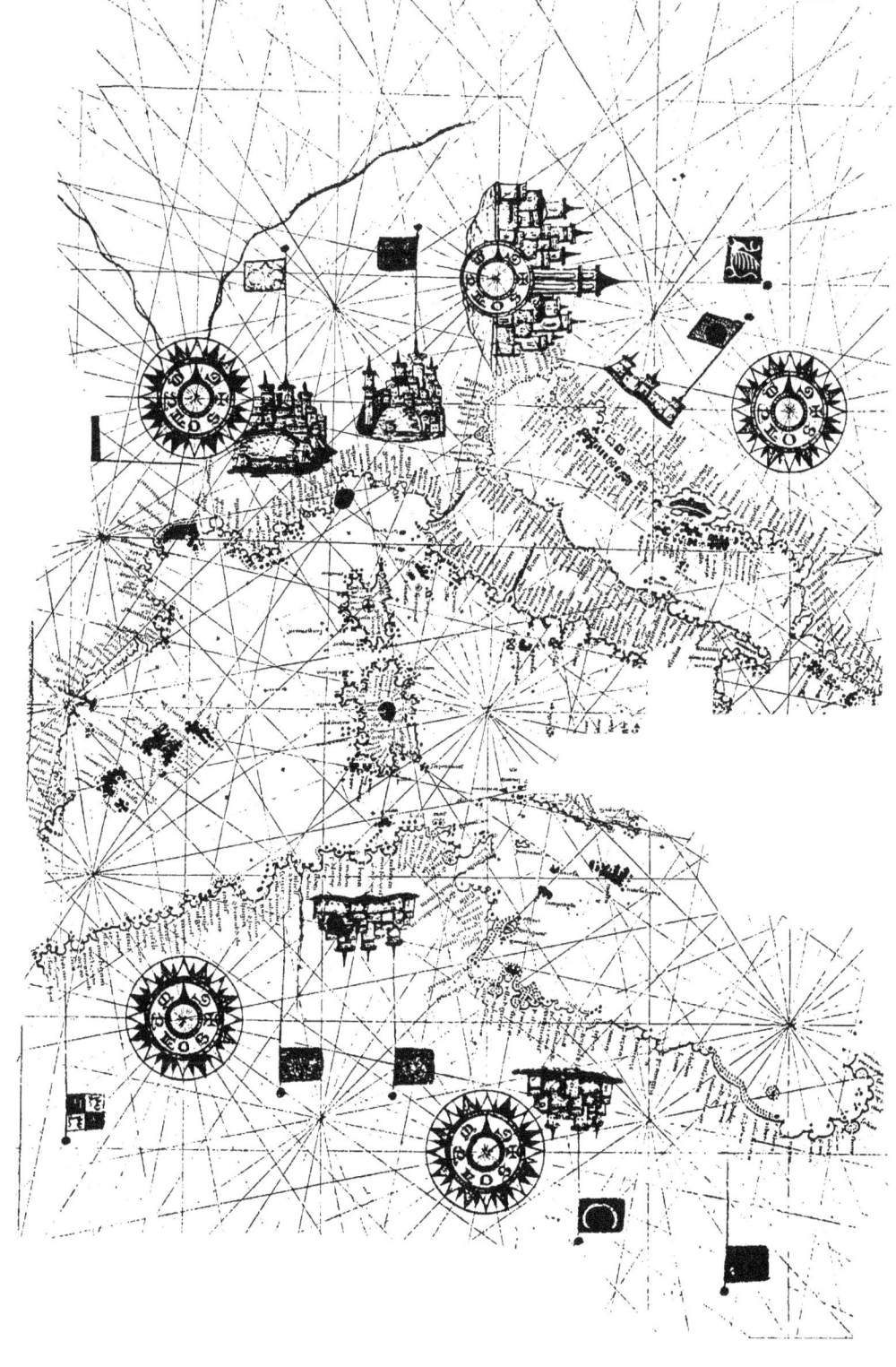

is highly colored, and contains numerous elaborate ornaments. " Europe," " Asie," and " Afrique " are appropriately inscribed, with the addition of such names as " Greece," " Troye," " Natolie," " Carmanie," " Svrie." Golgotha is represented with the three crosses. The mouth of the Nile is made prominent, as is the name " Barbarie."

4. Chart four includes the entire Mediterranean, with a small section of the southwest shore of the Black Sea, the northern section of the Red Sea, the Atlantic coast of Spain from Cape Finisterre southward, the coast of Africa to Cape Cantin. There is a very considerable tongue extension of the sheet on the left. It is rather more highly decorated than is either of the preceding sheets, containing the regional names and most of the features represented on each of the preceding charts. It appears indeed to be simply a chart representing the contents of the preceding grouped into one chart.

The entire atlas is one remarkably well preserved.

20. DOMINICUS DE VILLARROEL, 1590 circa.

An atlas of portolan charts, bound in a pasteboard cover and drawn near the close of the sixteenth century. It contains seven charts, each 37 x 52 cm. in size, with one page representing Judith and Holophernes having a Latin subscription concluding with a reference to the author: " Hoc opvs D. Dominicvs de Villarroel Regis Hispaniarvm Cosmography s faciebat," one page representing the martyrdom of St. Sebastien, and one page on which appear two circular calendar tables furnished with a movable parchment disc. Under each of the tables is an explanation as to its meaning and its use.

Villarroel was probably not a Spaniard, as he is not referred to by Fernandez de Navarrete in his *Bibliotheca maritima española,* but was probably an Italian living under Spanish rule in Naples. The Bibliothèque Nationale of Paris possesses a portolan chart, apparently the work of this same chart-maker, representing the Mediterranean, Europe, and northern Africa with the inscription, "Don Domingo de Villeroel, cosmographo de su Magestad, me fecit in civitate Neapolis 1589," and two atlases are referred to by Nordenskiöld in his *Periplus,* p. 65, of the years 1530 and 1580 which may be by the same cosmographer. This atlas, hitherto unknown, is probably his last work. Each chart is covered with the usual crossing lines and contains several compass roses, some of which are elaborately executed. Degrees of latitude are represented, on each also the scale of miles in a waving scroll. Certain important cities are made especially conspicuous on the first four charts, and numerous banners of state and coats-of-arms are represented in their appropriate localities.

1. Chart one represents the eastern Mediterranean and the Black Sea, recording in the interior regions only the names of the several countries.

2. Chart two includes the central and western Mediterranean, having compass roses which are especially well drawn. The several countries of central Europe are each distinguished by the representation of at least one city over which flies an appropriate banner.

3. Chart three includes western Europe from Gibraltar to the White Sea, and is remarkably well drawn. Spain gives excellent illustration of the statement (*vid.*

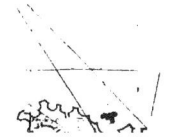

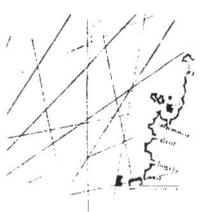

p. 20) that a different scale of measurement was used for the Atlantic coast from that used for the Mediterranean coast. England and Ireland are remarkably well represented as is the Baltic with the entire Scandinavian region.

4. Chart four includes southern Spain with the coast of Africa to the Gulf of Guinea. The prime meridian is represented passing west of the Canary Islands, on which meridian degrees of latitude are marked from 1° to 42° north.

5. Chart five presents the Atlantic Ocean from 15° to 60° north latitude. This is one of the most interesting charts of the Atlas, exhibiting on the right the coast of Africa, Portugal, and Ireland, at the top Iceland, Greenland, and the island of Frisland of the Zeno map, on the left Canada and Labrador with the neighboring islands under the name Terra Nova. In the middle Atlantic are many islands, among which are S. Brandan and Icaria. The North American coast represents a type between that of the Dieppe School and that of Ortelius as laid down in his Theatrum of 1570.

6. Chart six includes the Adriatic, Lower Italy, and Sicily. In this and the succeeding chart the draftsman has altered somewhat his style, giving less attention to ornamentation. The cities are made prominent merely by a gold dot. Compass roses are less conspicuous though in the style of the small roses on the preceding charts, and the ribbon scroll in which is represented the scale of miles is the same pattern.

7. Chart seven represents the entire Ægean Sea with the neighboring coast regions. As in the preceding chart ships are artistically sketched sailing the sea, and the towns, as well as all coast-places bearing name, are preceded by a gold dot.

A small portolan chart of the year 1597. Its dimensions are 17 x 54 cm. The author legend near the upper border on the left reads, "Vincentius prunes in civis majoricarum me fecit anno 1597."

Neither Uzielli e Amat nor Nordenskiöld refers to this chart-maker, although they make brief mention of Matteo Prunes as a Majorcan cartographer whose work belongs to the second half of the sixteenth century.

In the tongue extension on the left is a miniature of the Virgin and Child resting on a cloud, underneath which are the heads of three cherubs.

The sheet being so much longer than broad would apparently call for at least two systems of crossing points. Instead but one is represented with its centre in the island of Sardinia. The circumference of the circle in which the sixteen points appear passes through the Adriatic on the right and the east coast of Spain on the left. The lines passing through those points are extended to the borders of the sheet. Numerous parallel lines cross the sheet from north to south and from east to west at intervals of about five degrees.

Along the upper and also along the lower border a scale of miles is represented.

The chart includes the entire Mediterranean, also the Atlantic coast region, from Cape Finisterre to Cape Cantin, with an extensive nomenclature, but no interior regional names are given. Ten miniature representations of cities are drawn, over each of which is an appropriate banner.

The chart is well preserved, though evidently very slightly reduced from its original size by trimming.

22. ANONYMOUS ATLAS, second half of 16th century.

An atlas containing three portolan charts of the late sixteenth century, bound in brown boards. Each chart is 40 x 58 cm. in size. It is the work of an unknown French cartographer. The continental coasts and most of the islands are colored. The chief ornamentation consists of compass roses, each of which has eight points, the central one in charts two and three being located in the island of Sicily, and there are somewhat elaborate cartouches, in each of which a scale of miles is represented.

1. Chart one represents the Grecian Archipelago, including a section of the coast of Asia Minor, " Natolie," the island of " Candie " very prominent, the east coast of the Grecian peninsula, " Grece," and " Morea," and the coast of " Romanie." Chios and Rhodes are made conspicuous by means of their color and the silver cross.

2. Chart two represents the entire Mediterranean, the entrance to the Black Sea, the west coast of Spain, and the coast of Africa to Cape Cantin. Certain coast regions and islands have been made especially conspicuous by color. Regional names in large capital letters are inscribed, as " Spagne," " Europe," " Asie," " Barbarie," " Afrique." This sheet has the tongue extension on the left.

3. Chart three includes about the same as the preceding, except that nothing beyond the Strait of Gibraltar is represented, and it is drawn on a somewhat larger scale. Sicily is made the centre of the group of sixteen crossing points, although not in the centre of the sheet, and a second crossing point is indicated in the eastern

Mediterranean, which, however, is not represented as the centre of a system.

The atlas is well preserved in all its details.

23. ANONYMOUS ATLAS, close of the 16th century.

A portolan atlas containing three charts, belonging to the closing years of the sixteenth century, in size 35 x 62 cm. Though unsigned and undated it presents many features suggesting that it is the work of a member of the Oliva family. The drafting is exquisitely done, the decorations of each sheet showing workmanship of a superior quality.

1. Chart one represents the Ægean Sea with all of its neighboring coasts to the west, the north, and the east, with the island of Crete at the south. The coasts are colored with certain sections very conspicuous. It contains a scale of miles in an elaborate cartouche stretching entirely across the northern boundary, or as it appears in the atlas, on the extreme right. Compass roses are numerous, though not conspicuous, and the compass lines are rather more numerous than is usual by reason of the fact that each of the sixteen crossing points is connected with every other point, the lines being extended to the border of the chart.

2. Chart two includes the entire Mediterranean, with a very limited section of the southwest coast of the Black Sea, and terminates in the west at Gibraltar. Sixteen pictures of important cities appear. The Nile with its numerous branches is represented, as is Golgotha surmounted with the three crosses. The chart furnishes an excellent example of an attempt to represent along the

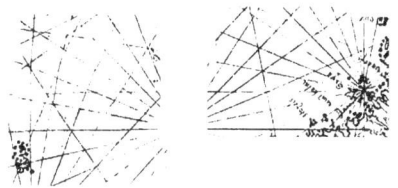

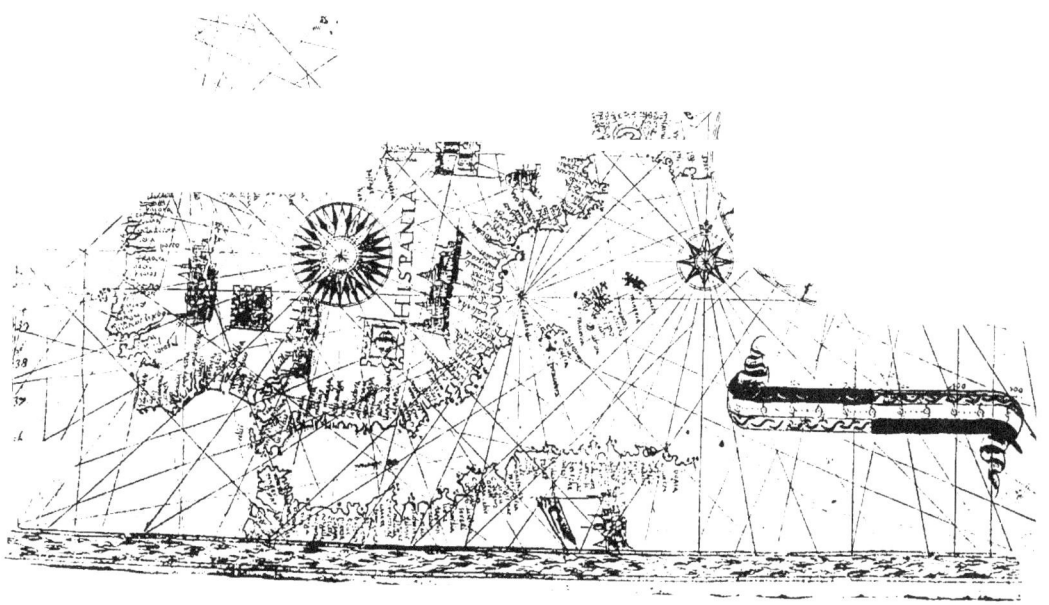

20. DOMINICUS DE VALLARROEL, CIRCA 1590.   CHART THREE OF ATLAS

coasts rocky promontories and sand shoals. In each of the four corners a scale of miles is drawn with a half border ' ornament.

3. Chart three represents the Atlantic coast of Europe and Africa from Holland to the mouth of the Senegal River, including the British Islands, the Azores, the Madeira, and the Canary Islands. Along the western, or as it appears on the chart as placed in the atlas, the upper border, the degrees of latitude are represented on a conspicuously drawn meridian line. Compass roses are numerous, and the crossing lines are arranged as on the other charts of the atlas. Two cities are represented in picture in France, three in Spain, and two in Africa. There is a mountain range in France running north and south and one similarly represented in the African desert with a spur extending to the southwest. A scale of miles is represented in the corner of the chart nearest Ireland, and one in an elaborate cartouche in Africa.

24. ANONYMOUS CHART, 16th century.

A small parchment sheet of the sixteenth century, 24 x 37 cm. in size.

It is clearly a double page chart from an atlas, having a narrow strip attached on the right to give the pages the requisite size.

In the upper corner on the right the name " Rocco Bagli " has been written, which is probably the name of a one-time owner. Compass lines and compass cards are represented with the usual system of crossing lines. A scale of miles is drawn in the upper corner on the right. Along the border of the sheet, and outside the colored

border, is a prominent representation of degrees of latitude, though numbers are not given.

Color was liberally used, especially for compass cards,
and along the coasts. Malta and Rhodes are covered
with the cross of the Knights of St. John, but no cities
are distinguished by miniatures, nor are flags or banners
represented.

The chart includes the eastern Mediterranean from
the island of Sicily. " Asie " across Asia Minor, and
" Barbarie " in northern Africa, are the only regional
names recorded.

The chart is not one of great importance. It is in
a fair state of preservation.

25. ANONYMOUS CHART, late 16th century.

A chart of the late sixteenth century, 18 x 19 cm.
in size. In the lower corner on the left is the name
" Bogali," probably that of a former possessor. It includes the Tyrrhenian Sea with the adjacent coasts and
islands, and has the usual crossing lines, with one compass rose. An ornamentation in the lower corner, near
the inscribed name, suggests that it is the work of a
French draftsman, though the nomenclature is Italian.
The chart is not that of a careful, expert workman, nor
is it one of great scientific value. Apparently it is a
sheet from an atlas, greatly reduced from its original size.
The names are all legible, though the sheet is considerably
water-stained.

26. VINCENTIUS DEMETRIUS VOLCIUS, 1600.

A portolan chart dated 1600, being 46 x 85 cm. in
size. In the upper corner on the left is the author

legend, "Viñus demetrei Volcius Rachuseus Fecit in terra Liburni de 13 Ianuari 1600." This chart is here first made known, but other portolan charts by Volcius which have been described bear dates 1593, 1596, 1598, 1601, and 1607. It includes the entire Mediterranean with the Atlantic coast from Cape Finisterre to Cape Bojador. The centre of the circle in which the sixteen crossing points appear is in southern Italy. There are five compass or wind roses, and on both the upper and the lower border a scale of miles has been inscribed.

The chart is remarkably clean and well preserved, having a considerable tongue extension on the left, and a narrow black border making an angle in the tongue extension, which border is omitted on the right.

27. MAIOLO E VISCONTE, 1605.

A portolan chart of the year 1605. In size it is 58 x 81 cm. In the tongue extension of the chart is a representation of the Virgin and Child, and the date 1605, which, however, has been crudely altered to 1505. To the right, and slightly below the picture of the Virgin, is the author legend, " Carta nauticatoria di mano de Baldasaro da Maiolo e Giouan Antonio de Visconte fatta nell' anno 1605 in Genoua," the year as here given having been also changed by the same hand.

Baldasaro was the last descendant of the famous Maiolo family of Genoa, especially distinguished as chart-makers in the sixteenth century. Apparently only two other charts of Baldasaro's are known, one of 1566, and one of 1583. Giovan Antonio was likewise a member of a famous family, of which family Pietro Vesconte was a member, whose name appears on the oldest-known portolan chart bearing date.

Compass cards are numerous, though not elaborate in design: a very simple scale of miles is represented on the upper and also on the lower border.

The chart represents the Mediterranean, the Black Sea, a small section of the Red Sea, the Atlantic coast of Europe from Gibraltar to Holland, and in faint outline a section of the Baltic and the southern extremity of Scandinavia, the coast of Africa to "rio doro," the British Islands, the Azores, the Madeira, the Canary Islands, and the fabulous islands "Maida" and "Brazil." Colored groups of turrets with flying banners represent the larger cities, among which Genoa, Venice, and Constantinople are especially conspicuous. In all eighteen cities are so distinguished. Much care seems to have been exercised with reference to the insertion of the British Islands.

In its colors and nomenclature the chart is well preserved, though the sheet has been slightly stained and torn in the margins.

28. JOANNES OLIVA, early 17th century.

Portolan chart of the first quarter of the seventeenth century, 51 x 96 cm. in size. In the upper corner on the left is a somewhat faded representation of Christ on the cross, to the right of which, and somewhat below it, is the inscription: "Joannes Oliva fecit in civitate Liburni año domini . . ."—the numbers representing the year having been erased.

The author was a member of the famous Oliva family, coming originally from the Balearic Islands, and later

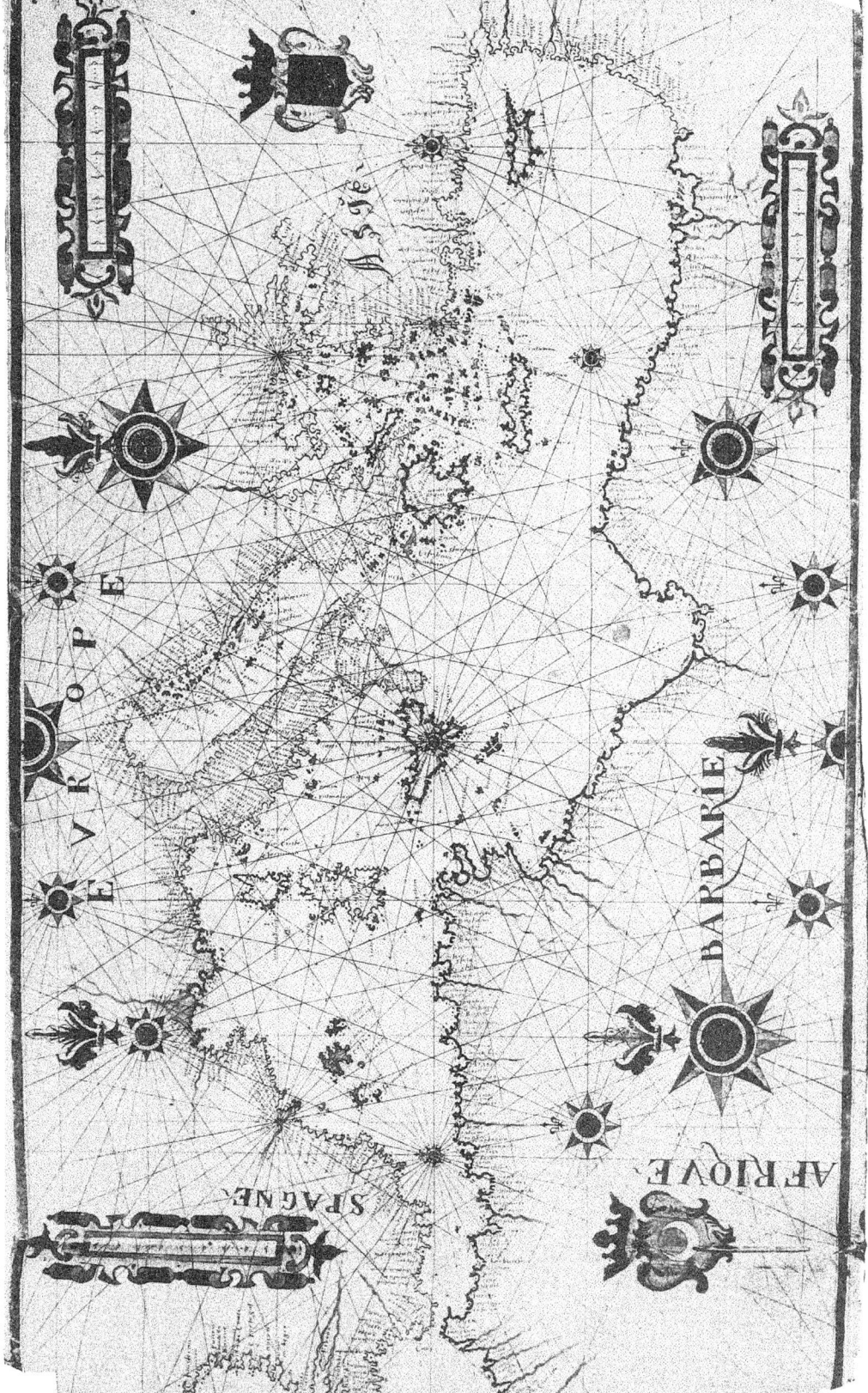

EVROPE

BARBARIE

ESPAGNE

AFRIQVE

having its representatives in many localities in Italy. Nine other single parchment charts and two atlases of his are known; the chart here described being apparently hitherto unknown.

Fifteen compass roses adorn the chart, though not all of the same design, the central one being located on the island of Sicily, from which the thirty-two lines radiate. A scale of miles appears on the upper border, and on the lower the scale is recorded three times in a long waving scroll. Latitude is represented on a meridian crossing the chart east of the heel of the boot of Italy. The chart represents the Mediterranean and the Black Sea with a small section of the Atlantic coast from Lisbon to Cape Cantin.

The three continents are designated by name. In Africa the river Nile is drawn, and in Palestine Golgotha with the three crosses.

The nomenclature is in Catalan, or in Italian with distinct Catalan forms, and as usual the names are in red and black. It is an interesting fact that those written in red are best preserved.

Over Rhodes, Chios, and Malta the cross is represented.

The chart is very well executed, but is somewhat injured along the edges, and in parts is slightly faded.

29. PLACITUS CALVIRO ET OLIVA, early 17th century.

A portolan chart of the early seventeenth century, in size 53 x 100 cm.

On the left in the tongue extension is the author legend, " Placitus Calviro et Olivia fecit in nobile urbe messana . . ." Numerous charts signed as here are known, possessing no superior scientific value, though well drawn and elaborately decorated. The date of this chart has been erased apparently with the thought of substituting another for the original. In the extension on the left is a miniature of the Madonna and Child. It is furnished with an elaborate border ornament, except on the right, where this appears to have been cut away. Degrees of latitude are represented on the left.

The chart includes the Mediterranean, the Black Sea, part of the North Sea, the Atlantic coast from Holland to a point near Cape Verde in Africa, with the British Islands, the Madeira and the Canary Islands. It is a chart elaborately ornamented, especially in Africa. West of Spain in the " Mare Oceano " two ocean monsters are represented. Various animals are to be seen in Africa, and palm trees are numerous. There are many miniature representations of cities, each with its appropriate banner. The three continental names have been inscribed, also Golgotha with the three crosses, and the Nile River, " Flume Nillo." No less than twelve rulers appear in their respective countries, each in full figure, with his appropriate shield. These represent " R. de Spania," " R. de Francia," " Imperator," " R. de Ungaria," " R. de Russia," " Gran Turc," " Gran soldano di Babilonia," " R. de Tripoli," " R. de Tunis," " R. de Alger," " R. de Fes," " R. de Maraco."

The purely geographical parts of the chart, including the coast lines and nomenclature, are very much faded, but the ornamentation is well preserved.

30. ANONYMOUS CHART, early 17th century.

A large parchment chart of the early seventeenth century, 77 x 93 cm. in size. It is unsigned but the general character of the draftsmanship and the brief inscription near the scale of miles reading, "Duytsche Mylen 15 voor een Graedt," suggests that it had its origin in the Netherlands though in the main the nomenclature is Portuguese.

As the chart represents the territory of especial interest to the Dutch West India Company, it is probable that it was drawn for use on one of its ships, and not long after the year 1621, in which year that company was founded. Later settlements in west Africa and Brazil are not represented. The map is covered with a network of compass lines in which the usual portolan chart colors appear, and one compass rose has been inscribed. Degrees of latitude are indicated which extend from 45° south to 43° north. On the right is represented the west coast of Portugal and Africa to the Cape of Good Hope, and on the left the east coast of South America from the mouth of the Amazon to the Rio de la Plata, North America and the West Indies being omitted. In the upper corner on the right is an inset map containing the west coast of Europe with the British Islands as far north as 60° north latitude. The interior regions are left blank, but the harbors, inlets, headlands, and mouths of rivers are represented, suggesting that the map was intended for practical use on shipboard. A stamp in the middle of the sheet indicates that the chart earlier belonged to the Dépôt des Cartes, Plans et Journaux de la Marine in Paris. On the inset map appear the regional names " Yrlandt," " Schotlandt," " Engelandt," " Francia," and " Hispania." In Africa appear the names "Marocho," "Mandinga," "Guinea,"' "Benin," " Loango," " Gout Cust," " Gabon," and " Angola," and

in the New World " Brasilia " and " Terra dos Patos."
The chart is remarkably well preserved and all names are
legible, the names of the Atlantic islands being in red.

31. JOUAN BATTISTA CAVALLINI, 1637.

A portolan chart, once part of an atlas, of the year
1637. It is 42 x 58 cm. in size.

In the upper corner on the left is the author legend,
" Jouan Batta Cauallini in Liuorno Año 1637." Caval-
lini was a chart-maker of distinction of the city of Leg-
horn, Italy. Uzielli e Amat mention three charts and two
atlases by this author, four of which are dated respectively,
1636, 1639, 1642, 1654, and one undated. It contains
numerous compass roses well executed, from each of which
thirty-two lines radiate. A scale of miles appears in each
corner of the sheet within an elaborately drawn design,
and the chart is furnished with an artistic border. Degrees of latitude are marked on a line crossing the sheet
to the west of the Madeira Islands.

The chart includes the western Mediterranean and the
Atlantic coast regions from northern France to " Arguin "
in Africa. The names inscribed for the larger areas are
" Spania," " Gallia," " Barbaria," " Africa." Numerous
cities are represented in miniature with banners, and near
each is the name of the province or region in which the
city lies. The provinces so distinguished are in Europe
" Piemonte," " Provenza," " Guascognia," " Navara,"
" Catalognia," " Valenzia," " Cartagena," " Andaluzia,"
" Castiglia," " Portogallo," " Galizia," " Biscaia "; in
Africa " Tunesi," " Costantina," " Algieri," " Oran,"
" Fesse," " Maroco," " Arguin."

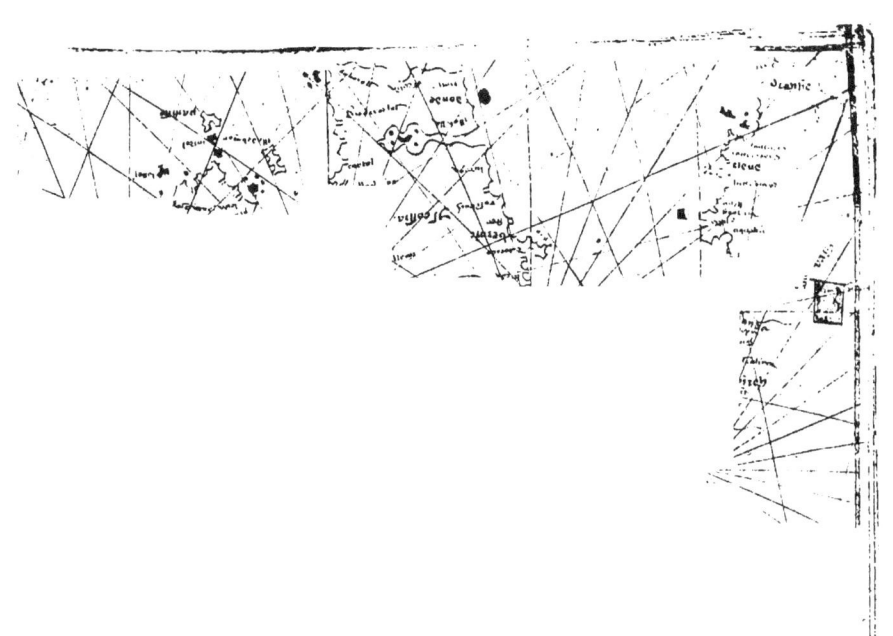

The elephant, the bear, and the unicorn are represented in Africa.

The chart is well preserved.

32. GEORG. ANDREA BÖCKLER, 1650.

An atlas of four portolan charts of the year 1650, bound in parchment cover, each chart being 29 x 42 cm. in size. In the lower corner on the right of chart one appears the legend, " Georg. Andrea Böckler, Archt. u. Ingeineur Ffort. 1650." This probably refers to the author: it may be an inscription of a one-time owner. We know that at Frankfort a/M. about the middle of the seventeenth century, there lived a distinguished architect and engineer by the name here given. Many of his works are extant, especially certain ones relating to architecture, to the science of war and heraldry, and there is also an engraved map of his known representing biblical history. The atlas appears to be a copy based wholly upon Italian sources of the fifteenth and sixteenth centuries, and closely resembles the work of the Oliva family. Each sheet contains compass roses with the usual sixteen crossing points arranged about a central point. We have here a fine illustration of the combination of wind and compass rose, the eight winds being designated in the central part of each compass card by appropriate initials, the north, however, having the needle point in place of the initial T, and the east the Greek cross, instead of the initial L. Each sheet has a scale of miles distinctly marked. Flags and banners are numerous on each chart, as are miniature representations of the important cities. The coast lines are composed of a series of short curves with numerous breaks, as if to indicate the mouths of streams or rivers.

1. Chart one represents the Atlantic coast of the old world from Cape Finisterre to a point near the mouth of the Senegal, including the Madeira and the Canary Islands. To the left of these islands a very conspicuous line is drawn on which degrees of latitude are indicated. " S.tiago " and " lisbona " are marked with turreted buildings and banners. Three cities, " Melli," " Ciudat de boxador," and " S. juan," are so distinguished in Africa.

2. Chart two contains the west coast of Europe and Africa from Holland to Cape Cantin, though the first point indicated at the north is " dansie ", with the Mediterranean coast as far as the meridian of Marseilles, including also the British Islands, and in the extreme northwest " Frixlandia," and to the south of this, " illa de brasill." The cities made prominent by colored miniatures include " frixa," " anvero," " Avignon," " barsalona," " valensia," " granada," " lisbona," " S.tiago."

3. Chart three includes the Mediterranean from the north coast of Spain to the Gulf of Corinth. On this chart Venice and Genoa are made especially prominent with picture and banner. Five other cities are indicated in a similar manner.

4. Chart four includes the eastern Mediterranean and the Black Sea. Banners are particularly numerous on this chart. " Gerussallem," " Mont de Sinayi," and " babelonia " (Cairo) are given special prominence. The Red Sea has not the usual solid red color, but is crossed with waving red lines.

The atlas is remarkably well preserved.

CPSIA information can be obtained
at www.ICGtesting.com
Printed in the USA
BVHW041429190219
540649BV00011B/107/P

9 781333 716141